D1580855

Ronald Giphart is an award-winn
strong interest in psychology and hu
Mark van Vugt when he worked as a c.
University Amsterdam

Mark van Vugt is Professor of Evolutionary, Work and Organizational Psychology at VU University, and research associate at Oxford University. He is the author of the international success *Leadership* (2011), and a regular contributor to national and international media channels such as the BBC, Channel Four, ABC, CNN, *Nature*, *New Scientist*, *The Times* and *The Daily Telegraph*. Van Vugt is also consulting editor of various psychology journals.

C334590568

Mismatch

How Our Stone Age Brain Deceives Us Every Day
(And What We Can Do About It)

Ronald Giphart and
Mark van Vugt

Translated by
Suzanne Heukensfeldt Jansen

ROBINSON

ROBINSON

First published in the Netherlands in 2016 by Uitgeverij Podium

First published in Great Britain in 2018 by Robinson
This paperback edition published in 2021 by Robinson

13 5 7 9 10 8 6 4 2

The publisher gratefully acknowledges the support of the
Dutch Foundation for Literature.

N **ederlands**
letterenfonds
dutch foundation
for literature

A CIP catalogue record for this book is available from the British Library.

ISBN: 978-1-47213-972-6

Typeset in Granjon by SX Composing DTP, Rayleigh, Essex
Printed and bound in Great Britain by Clays Ltd, Elcograf S.p.A.

Papers used by Robinson are from well-managed forests
and other responsible sources.

```
MIX
Paper from
responsible sources
FSC   FSC® C104740
www.fsc.org
```

Robinson
An imprint of
Little, Brown Book Group
Carmelite House
50 Victoria Embankment
London EC4Y 0DZ

An Hachette UK Company
www.hachette.co.uk

www.littlebrown.co.uk

Contents

The Clash

The idea for this book arose when a novelist was working as a writer-in-residence at the university of an evolutionary psychologist. They were preparing for a joint lecture about psychology and literature when the term 'mismatch' came up, a term that set their imagination alight. It led to hours, days and weeks of sparring, until it began to dawn on them that they should do something with this knowledge and these acquired insights.

What you are reading now are the fruits of a scientist's and writer's labour, a not immediately obvious combination. Science deals with the question of how and why the universe is the way it is. There may be many religious creation stories, but only fact can explain the origin of the entire universe, the Earth, animals, humans and all else surrounding it.

It is the task of science to fathom this factual basis. When all goes according to plan, scientists are guided solely by academic facts, or what are known as 'hard facts'. In the course of their enquiries they encounter endearing errors, the occasional fierce quarrel and dogged debate, yet all scientists have an overarching desire to know the truth, to substantiate the true story of everything.

Writers, conversely, have been creating their own wondrous and compelling truths for thousands of years. For them 'soft facts' are the starting point; they invent their own creation stories about the how and why of the world. Writers do not need to prove anything; they describe a universe that might have been. According to some people, not least the writers themselves, they get closer to the truth at times than many an academic.

In this book the scientist and the writer embark on a quest for one of our time's most evocative stories: the story of us humans, our influence on the planet and our clashes with its other occupants. We will present hard facts and theories, expressed through soft anecdotes and stories.

A hard fact is that our ancestors roamed the savannahs in Africa for millions of years. They lived in manageable groups of around one hundred to one hundred and fifty individuals. For two million years, our bipedal ancestors hunted prey and gathered nuts, seeds, fruit and honey. And then ... something changed.

In literature this is a called a 'volta' or turn, a turn that changes everything. The volta in the life of humanity was the discovery of agriculture. Over the past twelve thousand years, people have learned how to cultivate land for the production of food, on an increasingly big scale. The surplus calories which agriculture eventually produced allowed our ancestors to have more children, leading to a significant rise in the global population. People put down roots in settlements, which became villages and later still towns and cities, huge cities, which were many times larger than the groups of one hundred and fifty in which they had once roamed the savannah. Further voltas followed the advent of agriculture; the industrial revolution of the nineteenth century was superseded by the post-World War II digital revolution, icing the cake for humanity and the planet.

Since then, everything has changed; what we eat, where and how we live, how we dress, how our children are born and how we raise them, who our leaders are, how we engage with our environment, how we travel and communicate and how we try to recover from illness. People of today differ in infinitely more areas and ways from humans and anthropoids of earlier times. Or do they?

This book is about the dramatic clash between the first few millions of years of human history – the era that is called the Pleistocene or, in popular parlance, the Stone Age – and the past twelve thousand years following the advent of agriculture. Or, if we were to view human evolution as an hour, about the clash between the first fifty-nine minutes and forty-three seconds and the last seventeen hectic strokes. It is about the clash between our biological and cultural evolution.

What does it mean that our minds and bodies are adapted to life as hunter-gatherers on the open grasslands, while we now live in split-level maisonettes in a completely different social context? Has this rapid progress been good for us? How do we try to survive with our primitive brains, formed during the Stone Age, in a modern information society which changes fundamentally every ten years? In other words: how does it affect us when we live in centrally-heated homes with electricity for bread makers and tumble dryers, in cities housing huge numbers of people we do not know, and with living and eating patterns wholly unlike those of our ancestors?

In nature, organisms adapt to their environment. This happens through natural selection (later on in the first chapter we will explain how this works). Humans gradually acquired the ability to change the environment to suit themselves. Cultural innovations allow us to shape our living environment to our evolutionary needs for food, security and reproduction. We are able to withstand cold by building huts and wrapping ourselves in pieces of cloth sewn together; we negate bad vision by putting on glasses or inserting contact lenses; splint broken bones to support them while they are healing; suppress infections with drugs; and avail ourselves of IVF and Caesarean sections to have the children who would otherwise never have been born. Three cheers to us.

No other animal species masks and compensates as much for natural weaknesses through culture as humans. Cultural evolution happens much more quickly than the biological variety. A good idea can spread throughout the global population at a rate of knots – the smartphone, Netflix or Pharrell Williams' latest hit, for example – whereas it can take many, many generations and thus hundreds and thousands of years, for a genetic change to embed itself in the human DNA. Our ability to find quick solutions for intractable and unforeseen living conditions and to manage our environment enables us to live and survive anywhere on Earth. From inhospitable fjords in Norway to scorching deserts to the Isle of Wight. Thanks to our cultural abilities, humans rule this planet. Hip hip hurrah.

But in order not to sound too smug, there is a flip side. Not all the changes we introduce to our environment are good for us, whether in the short or long term. Sometimes a mismatch occurs, literally 'a wrong combination'. The notions 'match' and 'mismatch' are important evolutionary concepts. Biologists began to use the terms in the 1960s in order to classify the relationship between predators and prey within a changing climate. Take migratory birds, for instance, which return from Africa in the spring just as caterpillars are emerging and there is plenty to eat. Global warming means eggs hatching earlier and birds arriving too late to be able to enjoy their snack. Their migratory instinct no longer matches the emergence of caterpillars, resulting in a crash in the bird population.

We call it a mismatch when, as a result of a change in the environment the survival and reproductive chances of that particular species' individuals diminishes. Or in other words: mismatches arise when species are no longer well-adapted to their environment, as a result of which their survival and reproductive chances are impaired. A sudden natural event such as an earthquake or a volcanic eruption can cause such profound changes to the natural environment that the species is no longer able to live in it. Something of this kind happened to the dinosaurs some sixty-five million years ago. A meteorite impact (that, at any rate, is the current scientific view) made the Earth uninhabitable to such an extent that approximately half of the living animal species alive at the time gave up the ghost. Once the dust clouds had lifted, not one single land animal weighing more than five kilos appeared to have survived the aftermath of this intergalactic collision. They were unable to withstand the sudden change in their environment; a mismatch, in other words. Conversely, all kinds of small animals (such as mammals, reptiles and a small group of dinosaurs called Theropoda) were able to flourish as a result of the disappearance of their large competitors, a match in other words. The small mammals eventually evolved into elephants, lions, chimpanzees, whales and humans, and the Theropoda into birds. Viewed factually, dinosaurs as a family group have therefore not died out (at least according to the prevailing theory).

Mismatch can also arise when species themselves alter their own environment, resulting in them – and in their slipstream perhaps other species as well – no longer being well-adapted to this new, changed environment that they have created themselves. In this book, we will argue that humans, as a result of their evolved traits, have been able to intervene in their environment to such a degree that they have created countless mismatches for themselves and other species. The consequences of this are becoming ever more clearly and painfully visible. In modern society, we have to contend with a range of Western diseases, and biodiversity on Earth has declined dramatically due to our interference with in nature – both are consequences of mismatch.

Although mismatch can ultimately impact on the fate of entire species, in this book we are primarily interested in its effects on individual well-being, and especially that of humans. This is the domain of evolutionary psychology, the science that concerns itself with the evolution of our brain and behaviour. More about this later.

In evolutionary psychology we speak of a match when a person exhibits behaviour that serves his or her evolutionary interests, in other words, looks after his or her own survival and reproduction. Gathering calorific food and fleeing real danger are examples of evolutionary needs, as is finding a suitable partner with whom to start a family.

Mismatch occurs when humans exhibit behaviour that does not serve but instead damages their evolutionary interests. An example of this might be consuming too much calorific food, being afraid of non-existing threats, or falling for the wrong person. In this book, we will show that during those mere seventeen seconds since the advent of agriculture and, faster still, during the 0.3 seconds since the industrial revolution and the 0.03 seconds since the digital revolution, humans have altered their environment so fundamentally, that the likelihood of mismatch has increased in areas as diverse as nutrition, education, sexuality, work, politics, war and nature. We need to do something about this, and we would like to share these insights with you.

First of all, there are all kinds of scientific consequences to mismatch. Viewed through a mismatch perspective, all manner of social and technological trends are easier to understand, from the popularity of Facebook to the craving for cosmetic surgery and our attitude towards refugees. Secondly, mismatch impacts in numerous ways on our physical and mental well-being, our happiness. Understanding the mismatch theory enables us to do something about this, whether it be exercising more, choosing our leaders more wisely or feeling better at work or in our free time. Finally, mismatch means we may have to rethink politics and policy. Authorities, institutions and companies that are better informed about mismatch are better able to create an environment that matches human nature (and its capabilities and limitations). When, for instance, scientific research shows that people feel better in a largely natural environment

– the environment in which humans have evolved – should we not make places of work, school playgrounds and hospitals greener?

Getting to the bottom of mismatch and recognising its effects starts with gaining greater understanding about the evolution of this bipedal primate called *Homo sapiens*. Let's pack our bags and travel back millions of years in time.

CHAPTER 1

The mismatch vision

Eight million years ago, as a result of global climatological changes, a landscape came into being in several places on Earth that scientists call 'open grassland' or the savannah. Many mammals at the time adapted to this new environment: antelopes, hyenas, boars, horses and some primate species. Without open grassland, you and I would never have existed.

Roughly six million years ago somewhere – in what we call Africa these days – an ape species that was wandering around in the forests bordering these open fields split into two subspecies, probably as a result of geological circumstances isolating one group (temporarily) from the other. One of these subspecies became the present chimpanzees and bonobos, and the other – eventually – humans. The former stayed in the forest, the latter made for the savannah. This provided new opportunities, but not without a struggle.

Biologists and anthropologists have addressed the question as to what the landscape that created our species must have looked like. The African savannah largely resembled the Shire, the fictional landscape in Tolkien's *The Lord of the Rings*. The environment was more verdant than at present, there were more rivers and the average temperature was around twenty degrees Celsius. It is not for nothing our ideal thermostat setting at home; it is the temperature of the primordial environment, to which our body has adapted. The common factor was a rolling, green landscape with streams and ponds that we still appreciate today when we go on a camping holiday in the UK or France.

But hold on! Before we go any further ... Research into humankind's origin has been the subject of fierce debate for years. There is no field in which colleagues from different disciplines challenge each other as harshly over theories and interpretations as this one. We will not burn our fingers on this. But there is one thing everyone seems to agree on: over time, many branches have grown on the tree of humanity, some of which have broken off in the meantime. Archaeologists, palaeontologists, biologists and anthropologists know of 'pre-humans', with beautiful Latin names such as *Ardipithicus ramidus, Australopithecus africanus, Kenyanthropus platyops, Paranthropus robustus.*

We begin this book two and a half million years ago, when our ape-like ancestors developed a unique curve in the lower part of their spine which enabled them to walk fully upright whereupon their brains expanded considerably. From that point onwards scientists designated our ancestors as 'human', or 'homo', in all colours of the rainbow: *Homo habilis, Homo erectus, Homo ergaster, Homo heidelbergensis, Homo neanderthalensis, Homo naledi.*

A hundred thousand years ago there may have been as many as six (6!) different species of homo active, who did not always treat each other kindly. What's new? Even as recently as thirty thousand years ago on the Indonesian island of Flores, *Homo sapiens* lived alongside a *Homo erectus* called *Homo floresiensis*, nicknamed 'the Hobbit' because of its pygmean physique.

This we know: in the end only one species of homo survived, ourselves! The species *Homo sapiens* appeared roughly between two hundred and a hundred thousand years ago and became the eventual winner of the Human World Championship! Light the lanterns and pop the corks!

Anyhow, there are indications that a homo called *Homo ergaster* was able to use fire almost two million years ago. Perhaps early hominids lit fires to keep predators at bay and to prepare food. Later, *Homo erectus*, another distant human relative, was to discover how to hunt with fire, by driving herds of animals towards a cliff edge at night. At night, fires were kept burning.

The discovery of fire is the subject of many exhausting scientific squabbles. Some researchers put it later, mid-Stone Age, but when and however it came about, fire was more than a stove or protection against predators. More than anything, it saw to it that mankind was able to absorb food more efficiently by roasting it, without having to chew it for a long time – like a horse – and this in turn led to a decrease in our bowel size which gave our body more energy to develop our brain size. Over the past two and a half million years homo's brains tripled in size. What's more, from then on there was a central place for the group to gather and socialise: the campfire. No wonder that a log fire on a winter's night is still the acme of cosiness. Let's put some more logs on the fire . . .

Human life in prehistoric times

What did human life look like twelve thousand years ago? Let's give free reign to our scientifically sound imagination, using insights from global archaeological, genetic, psychological, neurological, cross-cultural and anthropological research amongst hunter-gatherer people such as the !Kung from Namibia and Botswana, the Hadza from Tanzania, the Tsimané from Bolivia and the Enga from Papua New Guinea.

We lived in groups, like practically all primate species. The gorilla, for instance, (with whom we shared a common ancestor around nine million years ago) lives in groups of approximately ten individuals. The chimpanzee (with whom, as previously mentioned, we shared a common ancestor six million years ago) lives in groups of thirty to fifty individuals. We topped them all: homo was the most social of all primates and lived in the largest groups.

Primeval human's social network could comprise around one hundred and fifty individuals; family, friends and acquaintances. This was the tribe. Obviously, members of the tribe were not within each other's sight all day; they spread out over a number of different camps within a large habitat. Each camp consisted of a number of extended families. Each extended family consisted of a family – man, woman, and a few children – augmented by grandparents, cousins, uncles and aunts. The extended family members would move from camp to camp, sometimes as much as eight times a year. They would follow pretty much the same route, to a spring or a place where family members would dwell. The members of the tribe would get together once a year to gossip, party and to find a partner – for the kind of things that happen during a night on the tiles or at festivals like Glastonbury or Reading.

Throughout their lives, from birth to death, our ancestors were surrounded by close and distant relatives pretty much all the time. They hunted, ate and slept together; as nomads, they literally went around together all their lives. Groups offered protection. A group allowed people to collaborate when they were hunting, picking and gathering. Sharing food was one of the most successful survival strategies. Going around together and sharing was a primeval variant on present-day life insurance, but one that would pay out directly during your life. We are sharers by nature.

Social relationships were extremely important. They were so important that we have something which no other primate has: *sclera*, or the white of our eyes. Proportionally, our eyes are surrounded by a large area of sclera. The advantage of this is that, in a group, you can immediately see what others are looking at. Other primate species concentrate more on which way heads are turned, we look at where eyes are focused.

Eye movement can reveal the state of social relations within our group. In a group, people look more at leaders than at followers. And if the group's leader looks at something, then the rest will do so as well, because there could be a threat of danger. Human babies follow the eye movement of their parents after approximately three months. If their mother looks to the right, then they will do likewise. This is the first, speechless form of communication. Language probably evolved only as recently as around three hundred thousand years ago. At roughly one year of age, babies engage in social 'coordination'. The mother looks at an object, the baby looks at this as well and then turns its gaze to its mother to confirm that they have both seen the same thing. This common orientation through eye movement is not matched by any other primate. We are the Looking Ape.

As previously explained, our ancestors lived in small, compact societies of around one hundred and fifty people. An extended family would criss-cross their part of the savannah, looking for areas with plentiful food, watering holes and camping spots. Decisions about where to head for next were made by consensus. Contrary to what history books tell us, our democracy did not begin in Athens or Rome, but on the African savannah. These societies were egalitarian, which means the nomadic groups knew no hierarchy, had few possessions and food was shared. The socialist Internationale ruled on the savannah!

But note: primeval society was by no means a hippy culture of 'anything goes'. The egalitarian ethos had to be closely guarded to give a group a chance of survival. If someone tried to swindle, cheat or dominate, he could be bumped off without mercy (witness anthropological research amongst present-day hunter-gatherer people and archaeological excavations in which a good many bashed skulls were found in mass graves). Common interests and mutually accepted, informal leadership held the group together. If you did not agree with something, you would move on to another camp on your own or with your family.

Scientists believe that every tribe had different kinds of informal leaders, without any one person actually being the boss. No presidents, directors or managers on the savannah, therefore. There was some division of roles based on talent, age and gender. There were hunters (for food), warriors (to protect against external threats), diplomats (to maintain good relations with other groups), peace-makers (who had to see to it that the group did not break up and that conflicts were resolved), organisers (allocators of food sources and group activity moderators) and teachers (who had to teach others in the group, especially the young, know-how, values and standards). But these specialisations were not absolute and everyone had to be able to do a bit of everything.

Daily life consisted largely of hunting, gathering and sharing out food. The most important division of labour in the group was between men and women, something we also see amongst our distant relatives the chimpanzees, gorillas and bonobos. Generally speaking, the men in the group kept themselves busy with hunting, and the women with picking and collecting berries, fruits and nuts. We deliberately write 'generally speaking' here, because obviously there were women who took part in hunting and men who stayed behind. It does suggest that gender roles are ancient, nevertheless, and this explains why the brains of men and women – despite a plethora of parallels – are made up and function differently in some areas (we hope we have formulated it carefully enough to avoid being chased out of the village with tar and feathers).

During most periods, food was scarce on the savannah. There were times of plenty, but they were often followed by droughts and food shortages. During these periods, human evolution could speed up very quickly. We are physically designed for a diet consumed by our ancestors, with fish, meat, raw vegetables, fruit, roots, seeds, nuts and honey – and not for a diet of Big Macs and Snickers on a bed of Walker's crisps. Our biological evolution has seen to it that in 'times of plenty' people try to absorb as many sugars, fat and other calorific foodstuffs as possible, as a buffer against times of shortage.

Our ancestors were dependent on hunting wild animals and gathering nuts, seeds, honey and berries to ingest calories. Obtaining this food required a large number of calories. Stalking a gazelle could take hours and success was not always guaranteed. What's more, once meat had been gathered, it was shared out with everyone who wanted some. As there were no fridges, meat could not be kept. The 'tolerated theft hypothesis' argues that sharing meat happened under duress: if someone had found food he was obliged to share it with others. If he did not, then his fellow tribesman would remove it by force. During prehistoric times thrift and stinginess were not appreciated if at the expense of the group. This is how we have arrived at the proposition that, as far as property is concerned, human nature is oriented more towards the political left than to the political right. Contrary to what most economists, bankers and political commentators think.

Family life in prehistoric times

Evolutionary anthropologists assume that in prehistoric times people had on average a few more children than in present-day western society: approximately four or five children per adult woman. But infant mortality was high, and in the end every couple raised on average just over two healthy children into adulthood. This led to a reasonably stable population for many generations, because these two children eventually outlived the two parents. There was no room for more children as there was not enough food. Mothers breastfed their children for four to six years and during this period they were not likely to conceive. Moreover, whilst trekking from

14

camp to camp, parents were only able to carry two children. So for a long time population growth was held in check. After the agricultural revolution the number of surviving children per household could exceed ten, which led to an enormous population explosion – with all the problems this entailed. We will come back to this in Chapter 2.

It goes without saying that in prehistoric times children were not educated in a school. As soon as children were able to walk, they were admitted to the 'children's group'. This group was a mixture of very young, young and almost adult children. As the group was relatively small in size, it did not contain many children of the same age. Grouping together a sizable number of children of the same age, as we often see in school forms and sports teams, is a novelty from an evolutionary perspective which can lead to excessive competition and is not conducive to children learning from each other.

Children were brought up by their parents, but for a significant part of the time also by other adults in the group, whereby everyone was generally less strict than in later times. In anthropology, this shared child care is called 'cooperative breeding'. There are present-day hunter-gatherer communities in which parents are barely involved in the raising of their own offspring. They do not stop their children making mistakes and do not protect them against a host of dangers, as they will have to find these out for themselves. It is important that children learn to recognise dangerous situations individually in order to survive in a hostile environment.

What were these dangers for our ancestors? Threats came from outside and inside the group. People were forever on the lookout for bloodthirsty predators and poisonous creepy-crawlies. At times people followed the same herd, at others they would forage in the same area. In areas with scarce water and food, clearly demarcated territories between the groups existed – but people did not always stick to these. Border conflicts ensued, involving a great deal of bloodshed: it is estimated that in prehistoric times up to 30 per cent of adult men died as a result of violence. But there was also room for peaceful relations between groups that had united into a large tribe and shared watering holes and camps with each other.

From within there was always the threat of scroungers who lived off the efforts and contributions of others, or of dominant individuals who tried to control the group. And finally there was the regular threat of violence between competing families. Family feuds as depicted in mafia films like *The Godfather* and *Goodfellas* had their origin in the African savannah.

Big Bang for humanity

Human hardware – our body and brain – was formed over a period of millions of years in a relatively orderly natural and social environment. Around twelve thousand years ago a fundamental change took place that turned the world on its head. In several locations on Earth, people began to settle permanently. Instead of wandering nomads they became farmers who cultivated their land and domesticated animals, perhaps because their hunter-gatherer ancestors had depleted the flora and fauna resources around them.

Hold on again! The question of where and how agriculture first came into being is fiercely contested in the scientific community. Most likely, more or less simultaneously, probably in several places at once – the Middle East, Central America and East Asia – groups of hunter-gatherers started to keep cattle and grow crops. The question is why this development happened so suddenly. A likely explanation is that around twelve thousand years ago the Earth's climate changed; the average temperature rose, making the circumstances for growing crops more favourable. To a much greater extent than before, it became possible to cultivate land and this is what happened. People began to settle alongside the crops they were able to grow, and kept animals which they domesticated rather than having to run after them for a meal. Agriculture is the 'Big Bang for humanity'. It was as if life was being reinvented. Groups of people who had moved around for centuries settled in one place in order to cultivate the land, and in doing so changed their environment for ever.

However, this did not happen without a struggle. Having examined human bones, archaeologists have found that the initial farmers were less healthy than the surrounding hunter-gatherers. The average farmer was smaller than his hunting and gathering peers. Clearly agriculture did not bring in much, at least not at first. Yet at some point it became more lucrative to farm than to chase after animals. Humans started to create food stores, stores that could be defended. Hunter-gatherers always had to contend with the problem that during their peregrinations across the land they were not able to bring very much with them, let alone fridges or IKEA cabinets. Some people would have squirrelled away a few things, but not on a large scale. Once humans became adept at growing crops and keeping cattle there was no going back. Most became farmers, and the groups of hunter-gatherers disappeared or became marginalised.

The opportunity to store food created a society that differed radically from the long history of nomadic hunter-gatherers. Its effect cannot be underestimated. Because it was possible to store grains, there was access to food throughout the year and this enabled more mouths to be fed. Women could have more children, and at closer intervals. Children no longer needed to travel or to be dragged around. Families grew.

This transformation happened more or less simultaneously from Mesopotamia to Mexico, as if people all suddenly asked each other why they needed to chase steaks on legs when they could just park them behind a fence. The advent of agriculture and animal husbandry engendered large-scale trade, as farmers produced far in excess of what they needed for their own use. People began to live in bigger groups and lingered in one place. The stores had to be protected against thieves and other tribes, and this led to settlements with lines of defence and armies to protect villages and land.

Leadership changed. A shift was made from informal leadership to a system in which leaders were officially appointed and whose duty it was to store food and distribute this during shortages. Diplomacy changed. It was the job of the chiefs, kings and emperors to guarantee peace with other settlements, but room for expansion arose, too. This led to wars and conflicts. People married. Men and women were officially joined in matrimony and in some places men were able to marry several women at once.

This agricultural revolution changed our physical and social environment fundamentally. In many cases the effects were positive for our descendants: the global population exploded. It is estimated there may have been five to eight million people on the planet at the beginning of the agricultural revolution. During the course of twelve thousand years this has increased a thousandfold. In that sense you could say there was a match: humans fashioned an environment in which they had more and more children. But all these changes also provoked a mismatch: our new environment confronted us with things which our body and brain have been forever playing catch-up with – and are sometimes at a loss with.

In the beginning there was Charles Darwin

Evolutionary mismatch occurs when species are faced with a rapidly changing environment to which their hardware and software – their body and mind – is not well-adapted. In order to understand mismatch we need to delve a little more deeply into evolution theory and the principle of natural selection, before we go on to pursue evolutionary psychology.

Before Charles Darwin, with his book *On the Origin of Species,* fired the starting shot in 1859 for the theory of evolution people assumed God had created the world and had placed life in a perfect 'match' with its environment: the giraffe amongst the high trees, the cactus in an arid zone. But since Darwin we have known that this notion is nonsense. Plant and animal species adapt to their environment by means of a long process of evolution through natural selection. How did Darwin think this worked? His theory – which he amended considerably throughout his life – is really very simple, even though far-reaching consequences are attached to it. Darwin started from three assumptions – all three scientific facts now – which together led to the conclusion that species have adapted to their environment.

First of all (fact 1): individuals are never alike. However much individuals within a species may resemble each other, each specimen differs in aptitude, appearance or behaviour. Let's take a classic example and introduce a giraffe living long ago called Gerald. He happened to have been born with an exceptionally long neck (not common for giraffes at that time). Raphael, who was born in the same group, had an ordinary, average-length neck, like all other giraffe calves. In other words, there is variation between individuals within a species.

Fact 2: Darwin assumed – without knowing anything about genes, the mechanism behind evolution – that the differences between individuals are hereditary, at least to an extent. This means that Gerald's children stood a greater chance of having a long neck than Raphael's children.

Darwin's third supposition (fact 3) was that there is always competition between individuals within a species, for instance when searching for food or a suitable sexual partner. Some individuals fare better in this competition in that they have particular external or behavioural traits that offer an advantage. Thus Gerald, with his longer neck, was able to reach leaves higher up the tree a little bit more easily than Raphael, which meant he had a little bit more to eat and was able to give his children a little bit more as well.

According to Darwin, these three suppositions (facts) – variation, heredity, selection – lead to some individuals being better able to produce offspring than others on account of the fact that they have particular traits. Their descendants inherit their parents' traits, and thus a genetic change spreads through a population and natural selection takes place. Through this slow, sticky process the species eventually adapts to its environment. The adjustments that ensure that individuals from a particular species are better able to survive, reproduce and look after descendants, are called 'adaptations'.

The purpose of these adaptations is not to allow the species as a whole to survive. This is a major misunderstanding in thinking about evolution. Plants and animals do not do particular things because they are beneficial for the survival of the species, but because they increase their own chance of survival. In thinking about evolution it was assumed for a while that if she-wolves suckled the puppies of another bitch, these she-wolves did this in order to ensure there would be enough wolf puppies for the species to continue. But we now know that this idea is not correct (more outspoken scientists will even say it is utter nonsense). The fact that a species grows in number is an incidental consequence of natural selection, but not an objective. Natural selection deals with competition between individuals, or sometimes between groups within one species, of which the best adapted traits ultimately remain. This is the idea behind the 'survival of the fittest', the term introduced by Darwin's contemporary Herbert Spencer, after he had read *On the Origin of Species*.

Back to Gerald and his cousin Raphael. Having a long neck gave Gerald so many advantages that he had more offspring than Raphael. Gerald's children and great-great-grandchildren all had long necks as well, which meant that in time only giraffes with long necks were born. A long neck is thus a physical adjustment (adaptation) to the everyday environment of the giraffe, the African savannah.

Some scientists now assume that the giraffe's long neck may have originated out of sexual motives, because giraffe males also use their necks to fight with other males to gain access to females. The one with the longer, stronger neck usually wins. Or maybe giraffes with long necks are a distinct turn-on for females. It would not be the first time that size mattered in the animal kingdom.

Darwin called this sexual selection. It is worth considering that natural selection comprises both an organism's adaptation to the natural environment (trees with leaves high up their branches, for instance) and 'social selection', adaptation to the social environment. For animals that live in groups the latter may be more important. One of Darwin's books (*The Descent of Man, and Selection in Relation to Sex* from 1871) specifically deals with this. Sexual selection is a specific form of social selection. It is about adaptations that make an individual more attractive to the opposite sex, so that they can be assured of offspring.

Female breasts are an example. They do not really have a survival function (if anything, they are a disadvantage in a nomadic existence). In order to give milk, a nipple and an internal milk reservoir suffices, as we can see from many other animal species. The human female's graceful, external rounded breast has a sexual function, however, as it makes women more attractive to men. Perhaps it is not surprising that in upright primates female breasts became important traits, as males could see how young and healthy a female was. Erect meant 'young', sagging meant 'old', and sufficient fat reserves meant that someone had sufficient calories at her disposal to be able to bear and feed children. Rounded breasts signified full, fresh storerooms! Breast enhancing surgery (usually enlargement or firming) can cost around five thousand dollars. And turns your female breasts into a peacock's tail. It will set you back, but it's of no use, other than to attract men. The mismatch is that it continues to excite us (we, too, plead guilty), even though nowadays there does not always appear to be a relationship between rounded breasts and fertility.

21

Evolutionary scientists differentiate between two types of sexual selection: intersexual selection (where a trait spreads because it makes someone more attractive to the opposite sex) and intrasexual selection (a trait that helps to win the competition with individuals of the same sex). The red deer stag's antlers are an example of a trait that has come into being as a result of intrasexual selection. The antlers can reach as high as seventy centimetres and are intended as a weapon in fights with other stags. The larger the antlers, the greater the chance of winning the fight, and the winner will gain exclusive access to the females.

There are, in short, various evolutionary avenues that can lead to biological adaptations of individuals and species to their environment. We must examine carefully how each trait has come into being. The giraffe's neck can be the consequence of 'pure natural' selection, but also of sexual selection. Combinations are equally possible and appear frequently in nature. First the long neck gave access to more juicy leaves, something female giraffes thoroughly approved of, as it meant plenty of food for their calves. Result: they fell en masse for giraffes like Gerald (intersexual). With their long necks the males could also easily eliminate competitors for their females (intrasexual). Whereas Gerald has long since died, a part of him lives on: his genes. In the long term evolution is not about the survival of individuals, but about the survival of the gene material we are made up of. Individuals are born and die, but their gene material continues to live on in their children and grandchildren (and so on). Since Richard Dawkins' book *The Selfish Gene* (1976) we have known that we have to focus on the genes. To borrow Bill Clinton's phrase about the economy, *it's the gene, stupid!*

Chance plays an important role in evolution. Gerald was born with a new or aberrant gene – a so-called mutation – or a new combination of genes which happened to have given him a longer neck than his cousin Raphael. Gerald fared better as a result, which meant that his long-neck gene was able to spread. In the end the long-neck gene spread throughout the giraffe population as it offered significant reproductive advantages. That's why now only individuals with long necks are born. Evolutionary scientists emphasise that natural selection is not a purposive process. There is no designer or god in action who thought up everything beforehand. Evolution is all about fortuity.

Gerald had the fortune that the long-neck gene offered an advantage. Some mutations are maladaptive, on the other hand, and then we are dealing with 'negative selection'. Maladadaptive traits tend to disappear from a population with great rapidity. Suppose Gerald's gene mutation had made his neck not only longer, but a great deal more flaccid. Then Gerald, despite his rounded belly, would have lost every battle for a female, not have had any progeny and the long-neck gene would not have been able to spread.

Sometimes there is 'neutral selection', when there is no clear advantage or disadvantage to a trait. Eye colour for instance. People with blue eyes fare as well as people with brown eyes, at least in terms of reproduction: they have on average roughly the same amount of children and as a result both variants continue to co-exist in the population. Eye colour is therefore called a 'selection-neutral trait'. The same applies to small, physical discomforts such as flat feet or protruding ears. As these features offer no advantages or disadvantages to procreation, they continue to exist within the human gene pool. Just like physical height: there are short and tall people. This means, again in all probability, that there are advantages and disadvantages attached to increasing height. Research shows that men and women of average height are considered the most attractive, and also have the most offspring.

It takes many generations before a trait is finally perceptible within the entire species. If Gerald and Raphael had lived in a forest with ubiquitous juicy leaves instead of on the savannah, Gerald's long neck would have offered fewer advantages than Raphael's. The natural selection of this trait would have evolved a great deal more slowly, or in all likelihood not taken place at all.

There is obviously a clear connection between the advantages a particular trait offers and the speed with which it spreads amongst the population. The greater the advantage, the quicker the natural selection process. A trait like 'look after your brood' instead of 'dump them unscrupulously when they are born' offers mammals an enormous advantage. That is why 'maternal care' probably spread so rapidly when it first emerged.

Darwin's evolution theory shows how, over a long period, a species can adapt perfectly to its environment, so that its body and behaviour enable it to live there. An animal or plant's appearance can often tell us something about the way its environment looks, too. An anteater is excellently adapted to an environment in which he can obtain food from termite hills and ants' nests. His long nose fits in the holes of these hills and his long, rolling tongue is sticky enough to swallow thousands of insects at once and swift enough to avoid bites. The penguin for his part is perfectly adapted to the South Pole. His wings were originally intended for flying, but flying no longer has any urgency in the remote, predator-barren Antarctic region. And so a penguin can use its wings for something else now: swimming in the sea and hunting for fish. From wing to fin: an evolutionary adaptation to survive.

These two examples demonstrate that adaptations are historical traits. They tell us something about the species' past, but not necessarily about the present or future. In the same way that the anteater no longer needs his long nose and tongue in the zoo, features that were beneficial in the past do not necessarily need to be functional once the organism's environment changes. Biological evolution is almost always overtaken by new developments, so in fact we are constantly looking at the past when we want to understand why humans and animals have particular traits. When the environment does not change appreciably in relation to the period in which this feature first appeared, we call this a match. The African savannah where Gerald lived, for instance, has been the same for millions of years. For his descendants, their long necks and the savannah are a match.

Of interest to this book are features that used to be functional in the past, but have ceased to be: mismatches. Evolutionary mismatch is an important concept, as it makes clear that evolution lags behind. Rapid change in the environment can land species in difficulty – sometimes to the point of extinction. The species' individuals take decisions that ultimately damage their evolutionary interests. Let's hope it will not come to that for us just yet (but we are nevertheless putting on the soundtrack of Wagner's *Götterdämmerung*).

The Costa Rican golden toad

Once upon a time there was a golden toad. This Costa Rican frog species was discovered by a biologist in 1966. For most of the year, the toad lived underground in the forest. He would only come up for a few days a year, in order to mate, and then only while it was foggy so that his golden skin would not be directly exposed to the scorching sun. Thanks to the miracle of nature he had babies happily ever after, until . . . the lumberjack clumped into the fairy tale in his oversize boots.

Since the middle of the twentieth century the fog has disappeared in entire sections of the rainforest as a result of local deforestation and global climate change. This had the effect of the sun roasting the golden toads alive when they surfaced to find a partner. The species promptly became extinct, partly thanks to our craving for fine hardwood garden furniture and luxury holidays. Consider this when you are next relaxing on your patio in your teak lounger.

The extinction of the golden toad is an example of mismatch

between a particular evolved trait (a 'breathing' skin enabling life below ground) which does not fare well in a changing natural environment. This mismatch affected individual toads, male and female, and ultimately had disastrous consequences for the entire species. There might have been a more positive outcome if an accidental gene-mutation had occurred which produced Costa Rican golden toads with thicker skins able to withstand sunlight – and if this had spread quickly throughout the entire population. They would have fared better than their fellow toads with a delicate skin, and after a sufficient number of generations the species would have been able to adapt to the changing environment. Unfortunately, as said, natural selection depends on chance and this scenario, as far as we know, has not materialised. The last golden toad was sighted in 1989, and in 1994 it was placed on the list of extinct animals. Exit the golden toad.

The culprit in this case was *Homo sapiens* who, by changing the planet, has created all kinds of mismatches for other animal and plant species. The dodo similarly found its nemesis. During the seventeenth century, hungry Dutch seamen arrived on the island of Mauritius, the only island where this species of bird lived. In the absence of predators, the dodo had not developed a fear of hunters, whereupon it came face to face with the most gruesome hunter of all: humankind. The mariners also carried rats on their ships, which had not featured on Mauritius previously. The rats went for the dodo's eggs, and this destroyed any chance of junior dodos. This double whammy landed the dodo (about which more in Chapter 8) in the history books once and for all.

Evolutionary psychology

When we look at the anatomy of the human body we look at our past. We have an anus (for 570 million years), eyes (450 million years) and we are able to differentiate colours (60 million years). Obviously, we share these traits with many other animal species we are related to genetically, to a greater or lesser degree. Since ancestral humans sought out the savannah, we have been walking more upright (for – probably – three million years), our mouths have become a lot smaller (two million years) and we have hardly any body hair (for – probably – around a hundred thousand years). These traits have evolved as they helped our ancestors survive in their natural environment. But how well are they helping us now, in an environment that has completely altered?

The impression might arise that the concept of mismatch is limited to physical traits, but this is by no means the case. Evolutionary psychology, a relatively new branch within psychology based on insights from evolutionary biology, presupposes that the brain evolved on account of natural selection. The brain too can be adapted better or worse to the natural environment in which it has to function. Over the past decades, developments in animal research, brain studies, genetics and cognitive sciences have fundamentally changed our understanding of the brain and its software, the mind. These developments support the core of evolutionary psychology: all behaviour is the product of the brain, which in turn is the product of biological evolution. It was long thought that our brain was simply a clever computer delivered with a clean hard disc at birth. But we now know that genes, DNA and heredity play an important role in how the brain functions and learns.

A core term in evolutionary psychology is EEA (environment of evolutionary adaptedness). Take your time to work this out. It refers to the environment in which our ancestors lived and to which – after a great deal of time – their bodies and brain adapted. The EEA of *Homo sapiens* is primarily the savannah on the African continent. On the plains of this grand habitat our ancestors, as a species, became increasingly well-adapted to their environment, that is, its flora and fauna. In short: our brains are equipped to live and survive in the wild.

The EEA of the human trait to 'walk upright' is likely to have begun around three to four million years ago when ancestral humans had settled on the savannah. Our physical frame is best-adapted to this environment and for that reason finds a sedentary existence problematic to this day. Likewise our human brain has an EEA that was largely formed in this same environment. So we can rightfully say that we have to try and survive with a Stone Age brain in a modern digital society.

The traits of the Stone Age brain

What does our Stone Age brain look like? We will elaborate on this later, but for now here is a short overview. Put simply, our prehistoric brain was focused on allowing our ancestors to live and survive in small groups of families and friends, who moved around in a natural environment full of danger. We can still see traces of this in how our modern brain functions. Our brain is particularly concerned with dangers we can perceive with our senses (so can see, smell, hear or feel) and we are relatively less quickly alarmed by dangers not perceptible to our senses (like climate change). In divisive issues our prehistoric brain tends to let our self-interest prevail over that of others or the population at large.

Additionally, the interests of family members, who are genetically related to us, weigh more heavily than those of genetic strangers. That's why we find it so difficult to be as empathetic towards refugees as towards family members in trouble.

Thirdly, our prehistoric brain is rather short-sighted. In the past, the future did not really matter, because you had to try and survive from day to day. Food could not be stored, so if you had a piece of meat you would eat it yourself and any leftovers you would share with your family. If you saw a beehive in a tree, you would take a little honey before another human or animal was able to climb the tree. This short-term thinking can be seen again in modern Western diseases such as obesity, which arises because people prefer to reward themselves in the short term (tasty bag of crisps) rather than in the long term (a healthy body).

Fourthly, the human brain is an excellent copier. To our ancestors, the advantage of living in a group was that they did not have to discover everything for themselves the whole time; they could copy the behaviour of others, for instance on how to hunt or light a fire. If someone ran away, you would do better to follow than to stand still. This propensity for conformity was useful in prehistoric times, but in a changing environment can lead to us adopting behaviour that ultimately does not serve our evolutionary interests, such as delaying having children because your hipster friends do.

Finally, our prehistoric brain is status-oriented, because during prehistoric times status meant a greater chance of having progeny. By working more, our status in society increases, but we forget to convert this into greater happiness, better relationships and more children.

Psychological mechanisms

In order to understand our behaviour better, we need to know a few things about evolution – but as luck would have it, we can turn to evolutionary psychology. Our thinking, our feelings are shaped by natural selection from prehistoric times. We have evolved psychological mechanisms that help us to deal with challenges in our environment that are – directly or indirectly – important for our survival and reproduction. Evolutionary psychologists make a distinction between the evolutionary function of behaviour and the mechanisms that produce the behaviour. The ultimate function of human and other species behaviour is reproduction, or more accurately, having as many descendants as possible who for their part can produce the next generation. But to achieve this evolutionary objective humans first have to achieve objectives closer to home:

people need to find a way to survive, acquire status, find a fertile partner; they need to be able to feed and raise their children.

In order to enable us to realise these goals, evolution has equipped the brain with a large number of psychological mechanisms. These mechanisms dictate what food we enjoy eating, for instance, which partners attract us sexually, but also which type of leader we want to follow and whom we can trust. Evolutionary psychologists assume that natural selection takes place at the level of psychological mechanisms, resulting in the best mechanisms within the population remaining (i.e. the mechanisms that ultimately lead to the most babies). These mechanisms are also called 'instincts', as they influence our perception, thoughts, feelings and actions, often at a subconscious level. These instincts respond to cues from our environment, which in turn influence our behaviour. Thus, our psychology has evolved for us to choose a partner with whom we will produce children. And that in turn determines whom we find attractive as a bed partner. In order to increase their chances of offspring it makes sense for men to fall for women who are able to have children, in other words, young fertile women. And this may be stating the obvious, but research shows that (do not keep us in suspense any longer) young women are considered to be the most sexually attractive!

These evolved psychological mechanisms could be regarded as simple heuristics or 'if-then' decision-making rules. So for example: 'If you want children, then choose a fertile woman to have sex with.' This sounds rather obvious. It figures that the decision-making rule 'if you want children, then choose a woman irrespective of whether she is young or old' is not so shrewd from an evolutionary perspective. Men who applied this rule during our evolutionary past had fewer offspring, as a result of which their genes slowly but surely disappeared from the gene pool. Men who applied a decision-making rule which served their evolutionary interests better, had more offspring.

Their sons applied the same rule because those rules are hereditary – and thus the sexual preference for young women spread through the male population. Using the logic of game theory, we can show how the evolution of these psychological mechanisms works.

Game theory

According to so-called game theory (a mathematical model developed during World War II and now applied to many science fields) when making a decision, people form lightning-fast, automatic judgments about what is the best option. For example: the majority of people who approach a red traffic light on a busy arterial road would not dream of ignoring this prohibition, because of the likelihood of accidents or spot-checking traffic cops. And yet there are always some idiots who do not take the slightest notice of traffic lights. At night, when it is much quieter, more people will deliberately jump a red light because the chance you will be involved in an accident is smaller, as is being caught by the police. Game theory is a model for finding out what people's strategic decisions are. These decisions are often not particularly rational, and evolution has seen to it that they usually lead to the desired (that is, evolutionarily most advantageous) behaviour.

There are many applications of 'game theory' and we will encounter some of these (the prisoner's dilemma, 'the tragedy of the commons' and political negotiations) later on. As far as evolution goes, game theory assumes that individuals use various strategies (read: decision-making rules) to survive and ensure progeny. These strategies, steered by the genes, compete with each other. The victorious strategy spreads over the population at the expense of strategies that are less successful.

It is possible to show that from an evolutionary perspective the decision-making rule 'only trust people you know well' is more successful than 'trust anyone, irrespective of how well you know them'.

The individuals who applied this latter rule ran the risk of being cheated, and if this happened often enough their genes would disappear from the gene pool. We will use the insights into game theory to explain why modern humans are predominantly (but not solely) monogamous; why some people are leaders and other followers; and why we all find it difficult to keep the planet clean.

The human brain is full of decision-making rules which were good for us in the environment in which we evolved (EEA). These 'if-then' rules were about partner and food choice, dealing with hazards, how to acquire status, whom to choose as leader or collaboration partner, what to talk about, etc. But we do not know if these psychological mechanisms and decision-making rules still lead to adaptive choices in the present or future. Our environment, partly through our doing, has been radically altered and so it is possible that the decision-making rules that were useful to us during ancestral times may now lead to the wrong choices.

For example: we are living in a world right now in which women (may) appear younger than their biological age because they wear make-up and have undergone plastic surgery. This might lead to men (with their decision-making rule 'if you want children, choose a young, fertile female') being tricked and falling in love with women who are less fertile or even infertile. As a result, these men's chance of progeny decreases. This is a clear example of a mismatch, of a pre-historic instinct that worked in the ancestral environment when the age of a woman was easy and reliable to assess. In present times this is not always the case. Because of a mismatch, our ancestral instincts do not always respond equally adequately to our environment.

Four irresistible cues

The decision-making rules in our brain are activated by cues from their surroundings. A cool glass of juice may tempt you to have a drink, for instance. Feeling hungry is a cue for eating, just as it is, for some people, for a visit to the supermarket. Mismatch-lesson: never go to the supermarket when you are hungry, because you will buy too much! Research shows that the human brain only pays attention to one or at most a few eye-catching cues in order to make behavioural choices. We call these the 'irresistible cues'. A mismatch may occur when our brain reacts to an irresistible cue, or when the correct cue to which our brain should respond is absent. These are the four most eye-catching mismatch cues, 'the four irresistible cues', or the

1 Exaggerated cues,
2 Fake cues,
3 Obsolete cues,
4 Absent cues.

We will begin with 'exaggerated'. The philosopher Plato asserted that 'beauty' is a manifestation of 'the good', but is this still the case? Take the cues that reinforce a basic instinct, but ultimately work against us. Our system has a preference for sweet things, as it evolved to encourage us to eat more of the less prevalent honey and ripe fruit, ingredients that contain essential sugars. Modern confectionery is much, much sweeter and it therefore has 'exaggerated' allure. Colourful confectionery is an exaggerated cue. We crave it, we eat it in large quantities, but it gives us tooth decay. In biology, this is also called a 'supernormal' cue.

This term was coined by the Dutchman Niko Tinbergen, winner of the Nobel Prize for Physiology or Medicine. He discovered that if an animal is offered a cue which its brain is instinctively geared towards, it will engage with it obsessively. In an experiment, Tinbergen placed a fake egg in a bird's nest. This fake egg was bigger and more brightly coloured than her own eggs. The bird focused all her attention on it and forgot to hatch her own eggs. In an exaggerated cue the most important characteristics of the original object of the instinct are magnified.

In addition, there are 'fake' cues, which look like cues from our ancestral environment, but mimic reality and therefore play a trick on our brain. These cues manipulate us to make choices that are adaptive in the real world and can lead to choices that do not serve our evolutionary interests. An example of this is internet pornography. When someone watches this (we, too, plead guilty), their brain and body react as if seeing real people in action, and the result is that men – like Onan in the Bible – spill their precious seed on the ground (or in an old sock). Around 10 per cent of people are addicted to internet pornography. From a biological point of view, porn-addicted men or women could make better use of their time and energy by finding a sexual partner in real life (more about this in Chapter 3, which deals with love and sexuality).

Then there are cues which were important to our ancestors, but have much less function in modern times. An example of an 'obsolete' cue is choosing a big, aggressive leader to defend your group. In prehistoric times it was important to follow a physically strong leader who could protect you against predators and hostile groups. But nowadays the physical strength of a leader is of much less consequence, as the leader has to be able to make the right decisions about often complex issues from behind his or her desk. So the choice for a strong, merciless leader has become less relevant in our new environment, and at times this choice can even work against our evolutionary interests. Just look at the millions of people who blindly followed Mao or Stalin, but ended up being victims of these bloodthirsty dictators.

Finally, in our modern environment some cues are entirely 'absent' which were present in prehistoric times. This can lead to a mismatch. An example is an excess of weak social structures in contemporary society where we are surrounded by people we do not know (well). Our ancestors lived in small, close, mostly familial groups, whose members were well-acquainted with, dependent on and helped each other when necessary. Let's compare this to the modern city, where the opposite tends to be the case. Not infrequently city dwellers hardly know their own neighbours or even not at all. These days, you can be lying dead in your house for weeks (or months or sometimes even years) without a neighbour noticing. Most people we encounter in the street or at work are genetic strangers. The absence of a close social network has many negative consequences for humanity, from social isolation and increased health risks to a rise in addictions and less happiness. Lonely people have higher blood pressure and suffer from raised cortisol-levels: the stress hormone that damages health. Lonely people end up living shorter lives.

In other words, exaggerated, fake, obsolete and absent cues can all cause a mismatch. Later we will attempt to chart these potential mismatches as accurately as possible in the sphere of health, education, work, religion, politics and the environment, in order to then hint at whether and how we can do something about it. Can we organise society in such a way that we respond to the right cues, without allowing ourselves to be misled? How can we turn a mismatch into a match . . . without necessarily having to go and live in a cave again or having to relieve ourselves without toilet paper?

More mismatch

Now that we have some understanding of evolutionary psychology, the terms match and mismatch are easily explained. Say you have a choice between two options, one of which gives you an evolutionary advantage (A) and the other does not (B). If you find A more attractive than B there is a 'match'. But say something changes in your environment or within you which makes you prefer B over A: then there is a mismatch.

In this book we are mainly concerned with evolutionary mismatches, that is, situations in which humans make decisions that harm their reproductive interests under the influence of changing environmental conditions. Yet, for humans it is equally plausible that they are confronted with cultural mismatch. Because one culture is not likely to be the same in terms of norms, habits and traditions as another, people may show behaviours that are inappropriate for the culture that they have moved into. For instance, a Spanish or Italian immigrant to the UK may find themselves in a situation where a long, two-hour hot lunch is not appreciated by their new employers. Or, a family from the South may be culturally mismatched to a life up North. In this book, we are primarily concerned with cultural mismatches to the extent that they are likely to have effects on the survival and reproductive chances of the individuals concerned.

Some examples would be: a child who prefers a bagful of colourful sweets to an apple (which may impact their health); an adult who finds earning money more important than starting a family; a moth who flies towards a patio heater because it thinks it is the moon; or a dragonfly who lays eggs on a windscreen because it thinks it is a reflecting water surface.

Likewise we are equipped by nature to run after our food, but meanwhile drive to the supermarket in our family car to load it up with fizzy drinks and ready meals. There is a global 'nutrition mismatch', with dramatic consequences such as diabetes and cardiovascular disease. The problem is that changing our taste is not something we can do in an instant. We happen to have an evolutionary preference for sweet and fatty foods, whatever good intentions we may have.

Another mismatch is our struggle with present-day leadership. The way we organise and manage things these days differs rather from that of our ancestors roaming in small groups across the savannah two and half million years ago. We live in countries with millions of inhabitants and work in big companies many of which have thousands of employees, and often we do not even know our boss. A large number of us live in a democracy, in which millions of citizens can take part in deciding who represents us. Are we choosing the right leaders with the right qualities? Does this new environment make the same demands on our present leaders as roughly a hundred thousand years ago?

A third example of mismatch lies in the genuine dangers that ambushed us in past times. We are still scared of spiders, even though European and American urbanised environments are no longer home to spiders that can cause harm. In the Netherlands, for instance, there are no venomous spiders. And yet many of us panic when we see a spider in the bathroom, and even seeing spiders on TV sends some people screaming. This fear instinct originates from a time when spiders did form a major threat to our ancestors in tropical climates, such as in Africa or Australia, where spider bites can be deadly to this day. But this fear response has no function whatsoever in a bathroom in Birmingham. We should replace our innate fear of spiders and snakes with fear of guns, cars, garden tools, wayward bars of soap, egg salad and incomprehensible modern kitchen equipment.

Plus and minus

Up till this point we have discussed examples in which the changing environment leads species to exhibit behaviours that are at variance with their genetic interest, but for the sake of completeness we should draw attention to the opposite phenomenon: that a vastly changing environment may give a particular animal species an enormous genetic boost, leading to a large number of offspring. We wish to differentiate between mismatches that damage the reproductive interest (minus matches) and mismatches that serve these interests (plus matches). The rabbit plague in Australia, for example, is what happened when a new animal ended up in a new environment without natural enemies. To the rabbit, this resulted in an almost unimaginable plus match; to many Australian plants and kangaroos a minus match. Let's go into this example a little more deeply.

It is the story of Thomas Austin, a British landowner in Winchelsea in the Australian state of Victoria. In 1859, he imported twenty-four British rabbits and a few hare, as he thought he might enjoy shooting them. It would make his estate in Australia look a little bit more like its British counterpart, he thought. But it is not for nothing that rabbits have a sexually tinged saying assigned to them.

During the first few years following their transit the twenty-four rabbits managed to breed as many as a million descendants. Sixty-five years after their arrival the creatures had become a population of ten billion, procreating with eighteen to thirty individuals per female rabbit per year (female rabbits can have offspring from as early as four months). One of the reasons that these animals were able to continue breeding without disturbance was because they had no natural enemies in Australia to make their life a misery.

It will not surprise anyone that the Australian landscape was seriously damaged by these newcomers. Everything was affected. The rabbits emptied the fields, nibbled on any seed they came across and in so doing destroyed areas that had been green for centuries. Thanks to these cute animals, original Australian animal and plant species got ensnared in an evolutionary trap. The Australian government came up with all kinds of measures to stop the plague, like a fence, the famous Rabbit-Proof Fence, that turned out to be far from rabbit-proof. More fences were put up, with an eventual length of 3,200 kilometres, without the rabbits taking the blindest bit of notice. Natural enemies of the rabbit were introduced (such as the fox), but they pounced on indigenous species as well, and that was not the intention. Initially successful was the spread of the myxoma virus that caused myxomatosis in rabbits. Within two years the rabbit population fell by almost 90 per cent. But soon natural selection began to manifest itself: rabbits with new gene mutations turned out not to be affected by the virus and were able to spread at breakneck speed again. Thus the import of twenty-four cute animals caused a great deal of trouble. British landowner Thomas Austin had a short-sighted goal in mind, leading to a permanent change in the entire continent.

There are other animal species where human intervention led to greater opportunities to reproduce successfully. Pets are the best-known example. Where would cats and dogs be without us? Domesticated animals (and plants) often look quite different from the animals (and plants) from which they originate. Around fifteen thousand years ago some wolves began to hang around people. The most friendly specimens were captured and subjected to a human regime. Over the course of a few generations these animals adapted themselves to us, and their genes were able to spread. The people who kept these tame wolves ensured for their part that their favourable traits were bred on. Wolves 'humanised' and became the dogs we know today; domesticated, social and faithful. In evolutionary terms, this happened in a flash. The Russian geneticist Dmitry Konstantinovich Belyayev managed to train wild foxes into sweet, tame animals in just forty years, an achievement that was described by the *New York Times* as 'arguably the most extraordinary breeding experiment ever conducted'. Many people considered Belyayev as the man responsible for the insight into how, in a short time, the wolf could be transformed into the loyal quadruped which can be found in our homes in its various guises, from Great Dane to Chihuahua.

This book focuses on negative mismatches (or minus matches), as these are the most important to us. We are happy to endorse the proposition that cultural changes following the agricultural revolution have yielded many plus matches for our species: after all, there are more of us than ever. We have more to eat, better hygiene and greater life expectancy. But does that mean a better, healthier and happier life for every individual on the planet? In his book *Sapiens* the Israeli writer and historian Yuval Noah Harari calls the agricultural revolution 'history's biggest fraud' and shows that the average person's life has not necessarily improved since the agricultural revolution. We attribute this to our primeval mind, which has lost its way in a new world of anonymous cities, social inequality and stress at work. Plus matches also carry a risk as, having ancient brains and minds, we often do not know how to cope in a world of luxury, comfort and plenty.

If we presuppose that our changing environment entails all kinds of risks and dangers, then it is best to focus our attention on the minus matches and to try and turn these into 'matches', in which our brain does in fact make the right choices that ultimately serve our genetic interests.

The cultural revolution

Our current living environment is completely different from that in which our brain evolved. And yet this has not resulted in humanity dying out. We have an advantage over most animal and plant species, which means we may escape the fate of the Costa Rica golden toad and the dodo from Mauritius: humans have culture. With the term 'culture' we do not mean Verdi's operas, van Gogh's paintings or the unread James Joyce novels on our shelves. These are just the tip of the cultural iceberg. Formulated more elaborately, culture relates to the ideas, norms and values which settle in one human brain and then alight in others through observation, learning behaviour and social interaction.

As we saw earlier, genes pass on evolutionary information. There is also a non-material version of genes: memes. The difference between memes and genes is that heredity of memes does not happen via DNA, from parent to child, but from brain to brain, via social interaction. This is a rather woolly way of saying that culture passes from human being to human being. By talking and looking, parents can impart things to their children about the world in which they live. In scientific language this is called 'vertical transmission'. When friends and lovers pass on cultural entities to each other scientists talk of 'horizontal transmission'; from teacher to pupil this is called 'diagonal transmission'.

This is not just a language game. Transmitting culture has proved to be all-decisive. Take the sheer fact that you are reading this book. We live in houses formed by culture, dress in culture, drive in culture, eat culture and even regularly make love in a cultural way (when we blindfold each other or send sexually tainted text messages).

Culture evolves through the principles of variation, selection and heredity just as much as through organisms, it is just that cultural evolution happens a lot faster than biological evolution. A meme can spread incredibly quickly in a group and – thankfully – disappear just as quickly. Think of eye-catching news stories, man buns, Crazy Frog or those ugly Ugg boots.

Culture can develop amongst people because of two uniquely human skills. One is language. By talking to each other people are able to spread cultural innovations in a group in a flash. When language is absent, as in chimpanzees, our genetic cousins, it is virtually impossible to extract and transmit all relevant information.

The other skill is imitation. We humans are super-imitators, we over-imitate. We are constantly checking each other out – to a large extent subconsciously – to see what is happening, who is all the rage, who has come up with something new, what everyone is saying and thinking, what are good table manners, how others conduct themselves. Thanks to being such champion imitators it is very easy for us to adapt to a new cultural phenomenon. When we arrive off the ferry in Calais and we see people driving on the right, we automatically follow suit (if we then see signs on the road saying 'drive on the right', we know we have done the right thing). No chimpanzee would match that: driving into the traffic he would undoubtedly crash to his death (apes should get better at 'humaning', if they want to become as intelligent as their family members).

Japanese researchers once observed a female macaque cleaning her potatoes before eating them. The other monkeys cast a brief glance at this behaviour, but only the young in the group adopted it. The older macaques looked at it, but were too set in their eating habits to do anything with it. They simply could not envisage that there was a connection between washing a potato and enjoying the food more. Only the younger monkeys looked on with interest and some copied the behaviour, without fully understanding what the idea was (washing in water = cleaning).

This is precisely the difference between people and other ape species. We learn by seeing the underlying purpose of the behaviour. Washing potatoes = greater enjoyment. Many apes watch potatoes being cleaned and might be able to copy this, but they lack the ability to make the link with the desired result. In science, this is called 'cue enhancement' (imitating behaviour without seeing its purpose) and it is considered an inferior form of copying behaviour.

Amongst higher animal species there are small cultural differences between groups because of learned behaviour. In chimpanzees and dolphins, differences in hunting behaviour or food gathering techniques have been established between groups that are genetically related but live in different areas. Some chimpanzees, for instance, engage in 'ant-dipping', whereby they poke a stick in a termite hill or ants' nest to lick off the termites as if it was a lolly. Other chimpanzee groups do not do this. It appears some copying behaviours within groups occurs, but this does not always lead to major changes.

Copying behaviour with a recognised purpose and benefit is much more powerful. Only humans seem to be capable of this. A baby smiling at his mother when she smiles at him is copying behaviour and one of the first manifestations of culture, as the mother imparts information to the child through her facial expression. With regard to copying behaviour, four-year-old children are at the level of adult chimpanzees. Using language and imitation, humans can avoid a great deal of mismatch. An example would be making warm clothes to be able to survive in cold areas without fur.

Why do we have these abilities whereas other intelligent animals do not, or, not to the same degree at least? A possible explanation might be that copying others' behaviours allowed early humans to live and survive in different natural environments with diverse threats and opportunities. For instance, being able to imitate the skills of the best hunters and foragers would ensure that people could find food in many different climates, from the equator to the North Pole. In addition, imitation would make it easier to get some degree of uniformity within groups that was needed to differentiate the group from other groups (common clothing, for instance, to recognise one another as members of the same group – something we call fashion these days). Our language reinforced this ability, allowing us to adopt things from each other without us having to be in each other's fields of vision. We were, and are, able to differentiate our group from others by emphasising different regional accents (orcas, dolphins and ravens also have dialects with which they recognise each other, for that matter). Our language also enabled us to know what was happening in our social networks, who was involved with whom, who did not stand by their word, etc. Language and culture are inextricably linked. That's why, to this day, we delight in language virtuosi such as philosophers, comedians, orators and writers (ahem).

Genes and memes

How do genes and memes relate? In evolutionary science, there are different views on the complex interplay between biology and culture. First of all, there is the view that 'biology always keeps culture on a leash' (after evolutionary biologist Edward Wilson's famous statement that 'genes always keep culture on a leash'). In other words, biology is the determining factor. Ultimately, cultural changes that are detrimental to procreation will not get anywhere, as the individuals who display this behaviour die, and their gene material along with it.

About Communism, for instance, Wilson said that it was a 'good idea' but unfortunately for 'the wrong species'. With this, he argued that the cultural notion of Communism does not fit the psychology of humankind, but that it does fit the psychology of the ant (highly recommended in this context is the animation film *Antz* about the life of a simple worker ant who, as an individual, tries to extricate himself from his billions of conforming congeners). Most political systems ending in -ism, such as Communism, Fascism, Liberalism and Capitalism are a mismatch in our opinion, as their view of humankind is too one-dimensional.

Suicidal behaviour is another meme that will not catch on quickly in our gene material. Various places on Earth have been affected by so-called 'suicide cults', in which several people – often in small, closed communities – choose to take their own lives. Between 2007 and 2009, no less than twenty-five young people between the ages of thirteen and seventeen committed suicide (mostly by hanging themselves) in Bridgend, a small community in Wales. Copying behaviour was the most likely explanation for these suicide copycats (see Chapter 9). It is obvious that a meme of this kind – 'suicide is a good idea' – is detrimental to individual procreation, and for that reason it will not spread through the entire population. According to this theory, memes cannot cause the extinction of genes.

Secondly, there are theories that presuppose that biology and culture can operate relatively independently of each other. An example is our diet. Our partiality for fatty and sweet foods is innate.

This is beyond dispute: everywhere in the world children are born with the same preference and this is advantageous from an evolutionary point of view. But cultural factors determine what we enjoy eating as well. Whether we like or dislike spicy food depends on our culinary experiences when we were young and that in turn is determined by where we grew up. Another example is language: all humans are born with the ability to speak. Our linguistic capability is determined biologically. But whether someone speaks German, English or Hindi is entirely dependent on which part of the planet they live in. Essentially, these two issues have nothing to do with each other.

Researchers have shown that culture and biology operate independently of each other particularly when ideas and opinions are transferred between friends and peers. Parents ensure that the cultural ideas they pass on to their children are advantageous to the reproductive success of their offspring. That is why many parents bring up the apples of their eyes in a gender-stereotypical way (pink clothes for girls and blue for boys), as this will present the greatest number of advantages later on and hopefully will lead to the greatest number of offspring.

But when cultural information is passed on between peers (via 'horizontal transmission'), this tends to have a limited effect. Being up on the latest music trends and fashion gadgets does not ensure reproductive success. Young people can have an adverse influence on each other, for instance copying smoking, risky driving (think of James Dean), unprotected sex or – more dramatically – suicide in Wales.

Lastly there are models and theories that presuppose that genes and memes are interdependent and able to influence each other mutually. If culture reinforces biology then we talk of a match, but if cultural factors weaken particular biological traits, then there is a mismatch. It is well known that in cultures with relatively high male birth rates (as in India and China), this is the consequence of a cultural parental preference for boys, coupled with a proven genetic advantage for parents who are able to conceive boys. So the combination of biology (increased chance of a boy) with a cultural predilection (preference for boys), can lead to a sharp evolutionary change in a population in the ratio of men to women. The result is a mismatch. Scientists speculate about the consequences of this mismatch in the number of men versus women, as exists in China for example. It is expected that this will lead to an enormous status battle in which many men will compete with each other to impress a limited number of women with their wealth. Another expectation is that men who are not able to find a wife will end up going abroad to secure a bride. Bride kidnapping was one of the most important motives for waging war in earlier hunter-gatherer societies. Venus forbid that this should happen again on a major scale. A third possibility is that men will adapt to women's preference for caring and monogamous men. The numerous men will have to compete for the favours of a woman, after all.

The notion of biology and culture influencing each other also finds support in a piece of DNA called DRD4. This gene has a short and long DNA-variant (allele is the technical term). People with a short DRD4-variant exhibit different behaviour from people with the long variant. The short variant is associated with altruism, justice and conscientiousness. In a study in which people were able to share out money between themselves and an anonymous other, the short DRD4 carriers gave more. The long variant predicts impulsiveness, risky behaviour, curiosity and a hunger for adventure. People with this gene gave less to other people in the game and were less prepared to divide the money fairly. The long variant is more prevalent in cultures with many recent immigrants, like the US for example. And the further away from Africa, where we originate from, the more prevalent this variant. Hence the term 'adventure gene'.

The fact that the adventure gene is more prevalent in migration cultures could be the result of natural selection on adventurous and risky behaviour. In areas where humans have been living for a long time and are self-sufficient, adventurous behaviour offers fewer advantages. Researchers believe that the adventure gene arose around sixty thousand years ago, when our ancestors started to migrate from Africa towards the Middle East, Asia, and later northern Europe and America.

A recent study in Kenya shows that people with the adventure gene fare better in nomad groups (as far as health and nutrition is concerned), whereas people without this gene variant fare better in farming populations. As a wandering nomad you are constantly on the lookout for food and a place to sleep, and that is why it makes more sense to have a curious personality. Conversely, in farming groups, where there is sufficient food, being social and able to share is important. Research into the adventure gene demonstrates that cultural changes – in this case migration to a new area – affect our biology and the selection of particular genes. If you place someone with the adventure gene in an environment with insufficient change, challenge and mobility, then you are dealing with a mismatch.

Match and mismatch

It can go either way with culture. The first possibility is that a cultural development advances our reproductive success, or 'cancels out' the 'mismatch' and provides a match. An early example involved clothing. In Russia, archaeologists have found needles made from bone and ivory that were used thirty thousand years ago to sew hides together. Being able to make clothes from animal hides is a cultural innovation that enabled our ancestors in the northern regions of Europe, Asia and America to survive. Culture created many advantages for individuals and their offspring, by removing the mismatch between our body (used to temperatures of twenty degrees Celsius or higher on the savannah) and an environment with below freezing temperatures.

Another match is lactose tolerance. The domestication of wild animals put milk on our prehistoric menu. Milk turned out to be beneficial during the growing phase and hence to the survival of the individual. The problem was that no primeval human of advanced age proved able to break down the enzymes in milk, something that still applies to 75 per cent of the global population. At some point a random chance mutation occurred in the DNA of some individuals which gives adults the ability to process the enzymes in milk. In Asia and America this mutation did not get a (cultural) hold, but in Europe and northern Africa – where camels and cows were kept – the gene ended up in the gene pool. The mutant gene spread through the population with great rapidity, especially during times of food shortages and it became extremely beneficial to drink cow's and camel's milk.

There are also plenty of examples of cultural traditions that create a mismatch. A – scientifically investigated – mismatch is the taboo on eating particular types of food (such as meat) in a region of Congo, for instance. This means some groups consume fewer calories than required to feed themselves and their children adequately. This applies in particular to women. One of the reasons for this is that when women marry, they leave their family group to join their husband and his tribe. There, they maintain the eating habits they learned during their childhood, even if particular ingredients are now missing from their diet. They then pass on this food preference to their children. These women do not adapt to the group, resulting in malnutrition in themselves and their children. The cultural adage 'when in Rome do as the Romans do' would be better for them.

Another example of mismatch has been described in Jared Diamond's book *Collapse*. A Viking colony from Scandinavia settled in Greenland. The colony flourished for four hundred years until, during the fifteenth century, it disappeared during a little ice age. Bone examination indicates they suffered malnutrition. The reason they died out was related to the fact that they had been farmers in

Scandinavia, a way of living they continued in Greenland. Until it got colder. The farming Vikings had never learned to adapt to the hunter-gatherer culture of the Inuit, whom they considered barbarians because of their strange and un-Christian habits. As they did not copy the Inuit in their behaviour – even though this would have been better – they died out.

A third mismatch example is postnatal depression. The question is whether the big cultural changes that have occurred in modern history have led to an increased risk of this condition. Firstly, we can ascertain that 13 per cent of women globally experience depressive symptoms within three months following the birth of their child. We obviously do not know what this percentage was in prehistoric times, but it is significantly higher in western societies than in traditional ones. These symptoms have negative effects on the quality of the relationship between mother and child, and this can lead to health problems and even an increased mortality risk in children. According to researchers, the cause for postnatal depression is to be found partly in mismatches in climate, nutrition, physique and parental care. To begin with, many young mothers have a vitamin D deficiency because of the limited number of hours of sunshine; in the Netherlands this applies to 60 to 84 per cent of pregnant women, for instance. A lack of vitamin D increases the risk of infection, health problems and depression.

Next there is nutrition. The primeval diet largely consisted of (red) meat from wild animals and fish, which contain a great many fatty acids. These fatty acids are often lacking in the present-day diet of grains and farm-reared meat. Another risk factor is bottle-feeding instead of breastfeeding. In the US for instance, only 49 per cent of children are breastfed during the first six months of their lives. Breastfeeding releases hormones (such as oxytocin and prolactin) which has a positive effect on the mental condition of young mothers and improves the ability to manage stress.

Cultural factors play a role, too. In traditional societies, care was shared with family members who were nearby. In the western world a young mother often has to cope on her own, as her family lives some distance away. Scientists call this the 'Latina paradox' (because young Mexican mothers in the US suffer much less from postnatal depression than white mothers; Mexican mothers may be poorer, but their families have much greater involvement in the care of their babies).

Small notion, enormous implications

Our mismatch theory presupposes that mismatch occurs when and where the environment changes so rapidly, that the bodies and brains of the species' individuals no longer fit the new, altered environment. In a greatly changing environment, organisms respond irresistibly to cues that are not to their evolutionary advantage. In this book, we will see that the agricultural revolution – and subsequently the digital revolution – has altered our environment to such an extent that a range of exaggerated, fake, obsolete and absent cues mislead our brains. By interfering in his environment, humankind has created a mismatch for himself and often for other animal and plant species as well. Culture can either cancel out or reinforce mismatch, because it is not always clear whether all cultural innovations are good for us. Some turn out well (clothing, medical care) and others (cars, supermarkets, offices, nuclear weapons, the pill, fascism) badly.

Mismatch is a small notion with an astonishingly huge number of consequences. It goes without saying that this book is not able to cover all conceivable mismatch in areas as divergent as education, health, leadership, religion, work, media and sustainability: but it can discuss the most important ones – even if occasionally, where insufficient scientific data is available, we have to speculate.

Awareness of mismatch can help us feel happier. When we know how our brain works – and what its limitations are – we can take action to make the right choices, choices that serve our evolutionary interests. Would it not be better to take some more exercise or modify our diet to lead a healthy life? Should we not work a bit more on our friendships instead of our career? Would it not be better to spend some more time in the real instead of the virtual world? The mismatch theory offers insights into how we should organise our society to resist all these irresistible cues that are coming at us from all directions.

Should we do something with all this knowledge? The natural fallacy theory states that even because something is, it does not necessarily have to be. This is an important lesson of this book. We may be naturally suspicious towards strangers, but this does not mean it is morally right and that we should accept it. In this book, we could go as far as simply describing mismatch and leaving it at that. That will not do. If we do not take any action, things will run their course and we will have to look on as many people suffer and die from Western diseases, hate work and their bosses, and produce too few children to prevent our population ageing.

But we can also use the mismatch theory to ensure that we steer the behaviour of people so that a match will arise. This can be done by altering the environment and applying knowledge from technology, psychology, economy and the law. It can also be done by 'nudging'. Ensuring that people get sufficient exercise in the workplace would not take all that much; we could organise companies in such a way that they are a bit more like the egalitarian tribes of the past; and we could advise people on how to kick their internet porn-addiction. We could also ban the payment of bonuses and the possession of nuclear weapons, and make cars so noisy that everyone can hear them coming from afar. We can elect our politicians for what they stand for rather than on their appearance.

The mismatch theory gives us the power to offer new insights into the mysteries of human behaviour and the human brain. Should we go back to nature? Is our rapid technological advancement normal? Why are we seeking greater status these days, not wanting more children? Does pornography lead to less real sex? Are we coming to grief evolutionarily on account of our prosperity? Does the pill make choosing the right partner impossible? Why do we have an arms technology that can wipe out the entire global population in one strike? Does advertising lead to greater extravagance? Why are we more afraid of snakes than of cruise missiles? Does Facebook make us unhappy? In short: how do we survive our modern information society with a Stone Age brain?

CHAPTER 2

Old bodies, modern diseases

I t is a vista French captain De Bougainville must have seen in 1768; paradise on Earth. There is not a cloud in the sheer azure sky, the sea ripples gently on the coral-white beaches with lush palm trees swaying in the tropical breeze. In the background we see a typical Faarumai waterfall. It is a scene that Paul Gauguin, who spent his last years on Tahiti, would have painted to such effect. In the foreground stand two smiling, staggeringly beautiful bare-chested girls in grass skirts, with colourful flowers in their hair. This is the island James Cook talked about with so much passion, the island of the noble savages and beguiling, willing women whom Charles Darwin, when visiting on the *Beagle* in 1835, described as 'well-proportioned'.

The Western tourists savour the spectacle for a few minutes before putting the postcards back in their rack. They have been on the island for long enough to know that the photo on the card in no way matches reality. Tahiti's coral beaches are in fact black, not white, but this does not bother them. The white beaches have been laid out for tourists. The thing that strikes them more than anything is that the inhabitants of Tahiti do not resemble in the slightest the image depicted in the photos and myths of the Tahitian dream women. The women they see here on the island weigh at least forty kilos more than the idealised pictures in the travel guides. Like the men, for that matter, most of whom also burst out of their clothes.

A local photographer has told the tourists that the models he uses for his 'typically Tahitian scenes' do not hail from Tahiti, as the Tahitians observe an orthodox form of Christianity. The proverbial free love, of which James Cook also spoke, has long disappeared. The last thing present-day female Tahitians will do is stand in front of a camera half-naked. Standards have changed, and one of the reasons for this might be that the mariners who discovered the Bounty island, with its free sexual morals at the time, brought syphilis with them. This resulted in a veritable massacre of the islanders, unfamiliar as they were with the exotic disease.

The souvenir shop in which tourists stand looking at the idyllic postcards of naked girls with flower garlands in their hair, is right next door to one of the three big McDonald's in the capital Papeete (on the corner of Avenue du Général de Gaulle and Rue du Dr. Cassiau). Only 26,000 people live in this small city, but the many hamburger joints are packed every day and by no means exclusively with tourists. Perhaps that should read, emphatically not with tourists. Visitors to the island are tempted into local restaurants with native dishes such as raw fish in coconut milk (*poison cru*) or suckling pig roasted in a fire pit (*pua*), but the local population tuck into Croque McDos (a McDonald variant of the French *croque-monsier*), quarter-pounders with cheese, French fries and milkshakes.

Food from the fast food chains is cheap compared to other eating-places on the island. Until the arrival of Westerners, practically all food was home-cooked and the only oil used was coconut oil. The Tahitians caught the odd fish and cultivated their land; that was all they needed. There were no shops or markets. Villages and communities ensured that fish were caught, fruit collected and that the land was cultivated. Families and tribes worked together and shared food. According to scientists, Western diseases like obesity, diabetes, cardiovascular diseases and high blood pressure were virtually non-existent on the island.

Diet in earlier times was not especially calorific and this may be why the ideal of beauty amongst young women and men was what we now call 'well-rounded' or 'buxom'. Before Tahitian girls and boys reached adulthood, they (or a group of them) would be subjected to an ancient custom called 'ha'apori', which literally means 'to fatten'. Youngsters from higher social classes would be locked up in special fattening houses, where they were stuffed with lovely food eventually to emerge as fat, healthy and attractive as possible. As food supplies were restricted, this had no damaging effects.

But alas, in short succession arrived British explorers Samuel Wallis (1767) and James Cook (1769) and the French explorer Louis Antoine de Bougainville (1768), followed by a group of Spaniards. It was goodbye to peace and self-sufficiency. The Westerners brought their own cooking habits to the island, led by the French. Their diet became fattier and more calorific and their taste buds were stimulated to an exaggerated degree (one of the four irresistible cues). Western diseases appeared, the French brought escargots who went on the rampage in and amongst the local produce – and much on the island changed. When, two centuries later, an international airport was opened and food was imported from overseas matters got irreparably out of hand.

What did not change was the ancient beauty ideal amongst the young Tahitians. These days, Tahitian girls are the third heaviest in the world, with all the attendant health consequences (diabetes, cardiovascular diseases, obesity). A significant number of the youngsters on the island are morbidly fat. The Western tourists stroll past the branches of McDonald's and see the fat necks, potbellies and double chins of the Tahitians. The islanders have gone from prehistory straight into the world of fast food and mass consumption.

Prehistoric bodies

Now that we have some understanding of the concept of mismatch, we will apply it to what is externally most visible: our body. Our bodies give information about who we are and about how and where we have evolved. As with other ape species our eyes are located in our forehead, not to the side of our head. We have hands, but as members of the exclusive club of anthropoids we can do something special with them. We have 'opposable thumbs', or thumbs that can circle and move independently of the other fingers. The evolution of this has been extremely important for our species, because it enabled us to throw, grab and keep hold of things. Without thumbs we would never have become what we are now: thumbs up for our thumbs!

We can also move forward on two legs. There are other animal (ostrich, kangaroo, penguin) and ape species that are 'bipedal', but we are exceptionally good at it. Bipedalism is not a sexual preference, but means 'two-legged locomotion', and that is something quite different from walking upright every now and then. Some of our ancestors, like *Homo Australopithecus*, were able to move bipedically, but this does not mean that they walked upright, like *Homo sapiens*, who moves forward with his head atop his neck. The fact that we stride along with such pride is due to the fact that we adapted to life on the savannah instead of in the jungle. Standing on two legs we were not only able to cover greater distances, but it gave us a better overview of our environment and kept our hands free to hold on to objects and gesticulate (possibly the beginning of the evolution of our language capabilities).

What is also notable is that we are predominantly hairless. Amongst well over five thousand mammals who do possess fur, this is pretty striking. Needless to say, scientists – as ever – are tearing their hair about what caused this. The classic theory says that we lost our body hair in order to regulate our temperature and moisture. The savannah has less shade than the jungle, where our fellow-apes dwelled. Our fur created too much heat in the sun and so we lost it. As we have perspiratory glands over our entire body, even a tepid breeze over the plain would cool us down. It is said that we kept the hair on our head as protection against sunstroke. An attendant problem was – at least according to this theory – that we cooled off too much at night.

During the middle of the twentieth century the alternative aquatic ape theory was popular. According to this hypothesis, eight to six million years ago our ape-like ancestors lived in areas where they had to gather their food in standing water, which offered protection against cat-like predators. Just as hippopotamuses do not have any hair because it is not a good insulator in water, so our ancestors were believed to have lost their fur for this same reason. In addition, we have a subcutaneous layer of fat, like other mammals who forage in water. As yet, no convincing paleontological proof for this hypothesis has been found.

Ten years ago, evolutionary biologists came up with a third suggestion: our ancestors lost their hair as a remedy against parasites. Fur is an ideal living environment for lice, fleas, ticks and other insect trash. When our ancestors discovered fire and began to wear clothes to arm themselves against the nocturnal cold, a pelt was no longer needed. This had the concomitant benefit that the problems caused by parasites (including itchiness, malaria, insomnia and Lyme disease) decreased.

Whatever the cause of hair loss: another theory suggests that sexual selection began to play a part, which meant bodily baldness became a sexual signal. Men (and women) found women (and men) without fur more attractive as baldness denoted an absence of parasites, and therefore better health. Hairlessness stood for hale and parasite-free. Head hair was retained because of the sunstroke-prevention mentioned earlier and we may have kept pubic hair because of its pheromones and other sexually attractive aromatic substances dispersing qualities. Beard growth was said to have stayed as an advertisement for manliness. Research shows that men with strong beard growth do not only behave in a more manly way, but are also considered more virile by women. The reason women's bodies are generally less hairy than those of men is probably because men have a strong preference for youthfulness.

A question people always ask is: when did ancestral humans begin to wear clothes and sleep under blankets? Scientists initially estimated this to be around a hundred thousand years ago, but following research into the evolution of lice this has been revised. To humans, lice are immensely irritating, but to science they are very useful as they lend themselves extremely well to genetic research. Approximately 170,000 years ago something happened to a part of the population of human head lice, which up until then had only lived in the remaining head hair of humans.

A group of lice migrated from this hair to the body. As the human body is bald, the creepy-crawly had no business being there, but when humans began to cover themselves in pieces of leather and other forms of clothing this broadened the lice horizon. Molecular biologists at the University of Florida discovered a change in the DNA of the two groups of lice, a change that coincides with the last but one ice age, when the Earth became too cold for many of our fellow beings. Clothing allowed humans to explore areas in which they had not been able to survive before, and studying clothing lice has helped us to establish roughly when this was in human history.

When we look at the hunter-gatherers of a hundred thousand years ago, what stands out is that nowadays we tend to be smaller, lighter and physically weaker. Forty thousand years ago, so well before the agricultural revolution, the average European man measured 1.83 metres. With the advent of agriculture the height of the average European was just 1.62 metres, a diminution scientists put down to climate changes and the shift to farming life. The first farmers experimented with new crops that did not yet provide the requisite calories. Six hundred years ago the average European stood at 1.65 metres tall and today this is 1.75 and rising. The reason for the increase in height is said to be improvements in health care and nutrition, and it may also be related to a decline in inbreeding as a result of urbanisation and globalisation.

Our brain has shrunk in size compared to a hundred thousand years ago (when we carried around on average 1500 cc, twelve thousand years ago this was 1450 cc and today it is 1350 cc), as have our jaws. Which is why our children end up at the orthodontist's nowadays. Their mouths and jaws are too small to house the many teeth that were useful in history. They need to wear braces and have their wisdom teeth removed to counterbalance this mismatch.

Finally, a few words about our prehistoric bodies: it is not true that our ancestors died at a younger age than we do. Or rather: it is only true in part. The average life expectancy of the hunter-gatherer who lived some hundred thousand to fifteen thousand years ago was in the end a lot lower thanks to infant mortality, but there were seventy- and eighty-year-olds in a tribe, just as there are amongst us today. Primordial humans had to contend with periods of famine, spiders and insect bites, large predators, much internal violence and high baby and infant mortality. This pushed down average life expectancy. But a forefather or mother surviving all that could certainly be expected to reach the age of eighty.

Having children

Women gave birth in solitude, far away from the camp. Following birth, women were able to decide for themselves whether they would take the child back to the camp. During prehistory it was not unusual to leave unhealthy babies behind in the forest. Raising an unhealthy child carried so many evolutionary costs that the group accepted this. Infanticide is a taboo subject in modern society, but it is perfectly understandable when we look at the difficult circumstances in which our ancestors had to raise their children. The killing of one's own offspring occurs amongst a great many animal species, in specific circumstances, for example when the baby is sickly, or when the mother does not feel physically or mentally able to offer it a decent future.

In 2009, in her book *Mothers and Others*, the American anthropologist and primatologist Sarah Blaffer Hrdy launched the by now widespread 'cooperative breeding' hypothesis: the idea that human children, as opposed to offspring of other ape species, were not raised solely by their mother, but by several adults from the social group. In her book, Hrdy describes that this process was set in motion as soon as the baby was born. After birth, the newborn was handed from adult to adult, as still happens in many contemporary hunter-gatherer groups. That way, the baby would make eye contact with everyone who carried it and this was a survival strategy as eye contact greatly reinforces the bond between child and adult (an evolutionary legacy of this is that we recognise people in love from the way they lock their gaze lovingly). For many generations, babies who were able to bond adults to them had a head start, as they would be better cared for, protected and fed. That's why babies arouse so much empathy in us, even if they are someone else's. The more baby-like features an adult face has, the more sympathetic we appear to find someone. Just think of celebrities with a typical baby face like Selena Gomez, Ariana Grande or Jim Parsons (from *The Big Bang Theory*).

In the ancestral tribe everyone was involved in the upbringing and well-being of children. Learning happened in the context of play. People were tolerant towards children and there were few 'don'ts'. Instances amongst present-day hunter-gatherers have been recorded where children have engaged in dangerous antics (like playing with a knife by the camp fire) without being corrected. They would find out for themselves that you should not touch a cactus.

This parenting style is fairly functional in a world full of dangers, in which you tended to be left to your own devices. According to evolutionary psychologist Peter Gray (in his blog Freedom to Learn, on *Psychology Today*), hunter-gatherers do not have any agricultural metaphors in their language. 'In their world,' Gray writes, 'all the plants and animals are wild and free. Young plants and animals grow on their own, guided by internal forces, making their own decisions . . . And that is the general approach that hunter-gatherers take toward child-care and education.'

Adults created an environment in which children were able to learn, but they did not feel any need to manage or encourage this learning. They treated children as they would treat adults. According to Gray, human children know best what they need. Children were afforded a great deal of trust, and so there were no or few battles between adults and children. The Brazilian psychologist Yumi Gosso wrote that adult hunter-gatherers 'do not interfere with their children's lives. They never beat, scold or behave aggressively, physically or verbally, nor do they praise or keep track of their development.'

Babies were taken everywhere, carried in a cloth, so that they accompanied the mother when she gathered nuts and fruits. From the sling on their mother's back the children would look into the big wide world, as opposed to most babies of today, who are lying in a pram and have eye contact only with mum or dad. Mother and baby were never apart for very long, in part because babies were breastfed for as long as until the age of five. If a baby cried, the mother would respond immediately. Up until quite recently modern educationalists advised mothers to let their children cry, as children would only demand more attention (German research showed that a third of children who cried were neglected by their parents, yet there has never been any proof that children who are comforted become dependent or emotionally damaged later on). During primeval times, it is unlikely that crying babies were ignored. If a child was crying, it probably meant something was genuinely up. Current baby manuals recognise this and advise parents to pick up their child as often as they like during the initial months, as during this early phase babies cannot be over indulged.

One of the big mysteries of human biology is why we reach such an old age when women stop being able to have children from as early as around the age of forty-five. Other mammal species die as soon as their fertility begins to recede, but we carry on living for decades. First, there was the mother hypothesis which presupposed that our long life was advantageous as mothers were able to take care of their little ones for longer. Children with mothers who lived longer had an advantage over children whose mothers did not. The genetic contribution of mothers who lived longer was marginally greater than the contribution of the mothers who did not. In evolution things work out in such a way that if a particular trait has an advantage, however small, it will eventually spread out over the entire gene pool.

The mother hypothesis was followed by the grandmother hypothesis. Women who have passed the menopause are no longer able to have children themselves, but by helping to care for their grandchildren they indirectly contribute to their reproductive success. The 'grandma effect' has been demonstrated in various cultures. Research shows that women whose mothers are close-by reproduce more frequently and successfully. Families with a local grandma have more children, and the effect appears to be enhanced if the grandmother is relatively young. Grandchildren with grandmas aged sixty or younger are 12% more likely to survive their childhood than children without a grandmother, Canadian, Finnish and Ghanaian research shows. For grandmas aged sixty and over this is a mere 3 per cent compared to children without a grandma.

The paleo diet

Over the past few years bookshops have been swamped with books about the so-called 'paleo diet'; titles such as *The Paleo Diet*, *The Paleo Revolution*, *The Paleo Diet Cook Book*, *101 Paleo Diet Recipes*, *The New Evolution Diet*, *Cavemen Cuisine*, *NeanderThin*, *The Primeval Diet* and *Primal Body Primal Mind*. The classic book was *The Stone Age Diet* written in 1975 by the gastroenterologist (specialist in the field of disorders of the stomach and intestines) Walter L. Voegtlin. The idea is that present-day humans should eat like hunter-gatherers in former times. According to critical food archaeologist Christina Warinner it is the fastest growing diet in America.

It's quite a job establishing what our ancestors would have eaten during the Palaeolithic period of the Pleistocene, the period between two and a half million and ten thousand years ago, supposing this could be unearthed over such a large time span. The so-called paleo-diet for modern humans excludes food that came about as a result of the agricultural revolution, so milled grains and legumes, processed sugars and oils or dairy products and manufactured products from the food industry are completely out of the question. What remains is meat, fish, fruit, vegetables, seeds, nuts and honey. In what portions our ancestors composed their menu no one knows, nor what the ratios between the various ingredients were. The paleo diet may sound scientifically sound, but it is largely nonsense.

Primeval man was by no means an explicit carnivore, as some paleo dieticians want us to believe. Food archaeologist Christina Warinner calls it 'the meat myth'; the idea that our ancestors lived primarily on meat. More than that, present humans have no anatomical, physiological or genetic adaptations for life on a diet of purely meat, according to Christina Warinner. We do have adaptations for consuming plants, on the other hand. Carnivores, animal species that eat only meat and no plants, are able to produce vitamin C themselves, but humans cannot do this. We always depend on plants (or pills, these days) for our vitamin C. Likewise our molars and other teeth seem more adapted to a diet of plants and fruits than to a diet of only meat. Scientists suspect that animal food represented at most 30 per cent of the total calorie-intake of primeval man.

Another fabrication is that our ancestors did not eat any grains. Many people who espouse the paleo diet believe that people only began to do this after the agricultural revolution, but thirty thousand year old millstones and pestles have been found, much older in other words than the beginning of agriculture. Recent examination of fossilised dental plaque in primeval people (someone has to do it) shows that our ancestors had microfossil residue of plants and grains – especially barley – sticking to their teeth. Paleo prophets do have a point when they say that a diet rich in bread, milk and cheese, say a Dutch breakfast, did not exist in prehistoric times.

But it should be noted that evolution itself did not lie down on the sofa with a satisfied tummy either. During the past twelve thousand years notable changes in the genetic inheritance amongst some ethnic groups have emerged, changes that have ensured – as we have seen – that the people of Europe and northern Africa are able to break down lactose sugars in milk, resulting in dairy products having become a welcome addition to our diet. This also applies to the animals, vegetables, seeds and nuts we feed ourselves with. These, too, tend to look completely different from those of prehistoric times, which begs the question: is there anything we can eat from the paleo period at all? Warinner cites the example of the banana, a fruit that seed-propagated in earlier times. All the bananas we eat nowadays are cloned fruits without seeds. The modern banana has been selected by humans rather than by nature – just like practically all other vegetable and fruit species we buy at the greengrocer's. The billions of bacteria that help us digest food differ genetically from the bacteria that assisted our ancestors.

To conclude, let's look at what our ancestors did feed themselves with. There is no simple answer to this, because food is always linked to time and place. Primeval humans were masters in flexibility where filling their stomachs was concerned. It goes without saying that they ate exclusively local products and adjusted to the seasons and their environment. If supplies in a particular area ran out, they would move to a different area. If they found themselves in a territory with few plants, they would eat more meat, and vice versa. The menu of a hypothetical restaurant at the ancestral campfire would have consisted of meat from small game, some meat from big game, from which especially bone marrow and offal would be polished off with great relish. People regularly ate grasses, they sometimes went fishing, quite often collected crustaceans from the water, frequently picked primeval fruit and vegetables, pulled primeval roots out of the ground, gathered nuts and if a tribe discovered a bees' nest it was party time at Il Grotto.

How do ancestral times continue to affect us?

In our bodies we still see the traces of prehistoric times. Take the way we move about, which is patently adapted to the savannah environment and not to an urban one in which train, car and bicycle play such a central role. A body perfectly adapted to modern life would look quite different. Cheery articles in which people imagine the adaptations which humans might evolve circulate on the internet. Why are we not born with shoes on our feet, for instance, like hoofed animals? And why do we continue to have tiny toes, with which we cannot grasp anything and which do not seem to be any use apart from stubbing against table legs and doorsteps (thumbs down for our tiny toes).

We can also see traces of prehistoric times in the way our children play. They enjoy games of hide-and-seek, leaping over ditches, cowboys-and-Indians. Subconsciously they are still practising for a life on the savannah with important life lessons ('hide yourself from a predator without making any sound').

Boys and girls

We will not be able to escape the topic of boys and girls in this book. Despite great similarities, research shows there are striking differences between boys and girls which are to do with the different roles men and women had in ancestral society. Inspired by the work of the British evolutionary biologist Helena Cronin, we can summarise these differences using four traits. These four traits are: talents, tastes, temperaments and tails.

On average, boys score better in tasks which test spatial visualisation ability, for instance how an object is turned in a three-dimensional space (a car that has to be parked into a tight spot, for example). Girls for their part score better in verbal tasks ('give as many words as you can think of starting with the letter G'). These differences occur throughout the world, but the degree in which they manifest themselves varies from country to country. This may be to do with the way education is organised, whereby gender differences are either stimulated or suppressed. Moreover, we are dealing with average differences, because there are obviously plenty of girls who obtain high scores in maths.

In general, girls are slightly more focused on people, and boys on objects and abstract phenomena. The statement 'I am happy when I can do things that make other people happy' is endorsed by many more girls than boys. Male fascination for non-living objects can be seen in one of our distant relatives. When we ask Vervet monkeys to choose between different types of toys, the boys are much more likely to go for cars and the girls tend to prefer dolls.

And then there are the differences in temperament. On average, boys are slightly more competitive than girls and are more likely to take risks. They come up with the craziest things to compete with each other, and if they lose, they invent something else in which they can get the upper hand. The list is long: from cook-offs to the world moustache-growing championships, or the latest sport, extreme ironing in which men decide who can most beautifully iron a shirt in the most challenging environment, e.g. on a high cliff, or in a sailing boat on the open sea.

Finally, and this difference is extremely significant, there are boy and girl differences in the so-called tails, the extremes of all kinds of traits such as intelligence and risky behaviour. Even though the scores by men and women for the previous four Ts (talents, tastes, temperaments and tails) overlap greatly, what is striking is how the differences in the scores are distributed. This spread is much wider amongst boys than it is amongst girls. Take intelligence. Although men and women on average do not differ, amongst men you will find a relatively large number of extremely intelligent individuals, but also a great many very stupid ones. Amongst women this spread is less extreme. There are more girls than boys with an average intelligence. This last difference can probably be attributed to how sexual selection works in nature, whereby males compete with each other for the favour of females. The majority of women will have children anyway, men have to make more of an effort. That means competition amongst men is greater and this in turn creates greater variation in male traits.

The agricultural revolution

When people began to settle down in fixed places and keep animals, our eating and living pattern changed in a relatively short space of time. This had consequences for population size (which increased) and our physical size (which decreased). During the initial phase of the agricultural revolution the available number of calories dwindled, probably because the human diet became less varied and quite a few harvests would have failed.

Scholars cannot agree about the question as to why people continued to farm and did not simply go back to their nomadic existence. The first dozen or so generations of farmers did not eat well, and the average height soon shrunk (which is related to the potential genetic bandwidth: if you eat well, you will grow just this bit taller; if you eat badly, you will end up just this bit shorter). But although the food was less healthy, there was enough to have more children. Nomadic ancestors would have two children on average, but farmers could house many more in their place of settlement. And when the population grew, the call for better, more efficient agriculture intensified as well. Some farmers will have given up their farming existence, but others had invested in planting crops of grains and did not want to leave all that behind.

Once we had mastered agriculture, our lifestyle changed fundamentally. A few grains and dairy animals began to play a dominant role, not only in our diet. We settled permanently in their vicinity, which weakened our inclination to move. The result was that we had far less exercise than our hunting and gathering forebears. There were also consequences for parenting, as families became bigger. Children had to be given more formal training, in schools and workshops, in order to satisfy the more specialised requirements in the fast-growing settlements. In a short time we altered our existence 180 degrees, something that the industrial revolution exacerbated even more. It was our bodies that lost out.

Mismatch

Following agriculture's difficult initial phase our birth weight increased, with all the misery this entailed. Even in prehistoric times, childbirth was quite a business as we humans have big heads (because humans needed big brains to survive and thrive in large groups) and women a tight birth canal. It is called 'the obstetrical dilemma': if women had a wider pelvis this would be at the expense of their

68

physical coordination and mobility. And so babies leave their temporary hotel room through a very narrow canal. The agricultural revolution exacerbated this existing mismatch.

The phenomenon of fathers wishing to be present at their children's birth is another example of mismatch according to an obstetrics specialist; an evolutionary new phenomenon that only raises stress in the mother and aggravates the delivery. When one of our children was born in the UK, the advice was to go to the pub and wait there.

Human babies are vulnerable and require help. That's why they tend to stay close to their mother, who can let them feed when they need to. Throughout history and cultures babies slept with their mothers (and often with their fathers, too). This was common in western countries as well, until in the nineteenth century the cultural notion developed that sleeping with babies was unhealthy. It was based on the craziest theories: sleeping together with a baby would stimulate it sexually or increase its chances of becoming homosexual. Other fears were that an adult mother would fall asleep on top of her child and suffocate it. An utter mismatch.

The anthropologist James McKenna of the Mother–Baby Behavioral sleep Laboratory in Indiana conducted extensive research into the sleeping behaviour of babies and has refuted all recent cultural nonsense about the subject. His research shows that it is safe and good for babies to sleep with their mothers and fathers. Babies do not have a fully developed nervous system, which means they sometimes 'forget to breathe', in extreme cases resulting in cot death. When a baby sleeps next to its mother and wakes up every so often to suckle, it is less likely to forget to breathe. Babies imitate the breathing pattern of their mother or father, and that keeps them alert. When a parent turns over, this encourages the child to breathe.

Another feature of shared sleeping is that, thanks to night feeds being relatively problem-free, these babies ingest more calories than babies who sleep alone (and for whom in many cases a bottle has to be heated on the bottle warmer). The uptake of immune elements via breast milk is considerably greater than via bottled milk. Here, too, culture plays a part. In western Europe, during the eighteenth century, breast feeding began to be seen as the habit of poor people in some circles. People preferred to feed babies with watered down milk or even sugared water. Bottle feeding is an example of a fake cue to which the baby brain responds positively, but which ultimately has a negative outcome for both mother and child. A story about Wolfgang Amadeus Mozart suggests he was raised on sugared water as it was considered to be more civilised than breast milk. It may explain why he died young himself and why – he adopted the same custom – four of his six children died before they were age three.

Sugared water or diluted cow's milk cannot hold a candle to breast milk which contains many substances that are beneficial to humans, including lysozyme (anti-bacterial), lipase (kills parasites), growth factors (immune system programmer), antibodies, epidermal growth factors, interferon (kills viruses), interleukins and tumour necrosis factor (anti-inflammatories), lactoferrin (iron binding), prebiotics (food for good bacteria), nucleotides (promotes immune response), taurine (supports brain tissue), lactose (aids brain growth), etc. Breast milk contains twice as much lactose as milk from cows, which admittedly, do not have as many brains as us.

All health organisations, such as the World Health Organization, advise mothers to breastfeed, as the nutrition is free, easy to digest and always available, has the right temperature and is free from potential bacterial infections. For babies there are also many proven and many presumed benefits. Babies who breastfeed throw up less and suffer less from gastroenteritis, and there is thought to be a reduced risk of diabetes, obesity and allergies. Other advantages at a later age are lower blood pressure, better teeth and higher scores on intelligence tests. But it is not only the children who benefit, there are advantages to breastfeeding for the mother as well: the risk of various types of cancer reduces in women who have breastfed for longer than six months.

In the Netherlands, the numbers of women who breastfeed their children rose until recently. During the 1970s as few as 47 per cent of women breastfed their babies, in 2007 it was 81 per cent. Three years later this number had shrunk to just under 74 per cent, however, despite government and WHO advice. Nowadays it is mainly the older and highly-educated mothers who continue to breastfeed; amongst younger and less educated mothers, the numbers have dropped in recent years.

There are obviously many cases in which there was no other option but to give the babies powdered milk. No one should feel guilty about this, because present-day bottle feeds are a perfectly good alternative. One of the authors of this book drank exclusively from a bottle from the third day of his life due to his mother's health (and as a matter of fact, still does). Apart from being an asthmatic, allergic, slack-shouldered, degenerate, deskbound slouch, it made little difference to him. Still, we should not think too lightly about bottle-feeding. We have adapted to our modern living environment with the help of culture. A cultural innovation is giving babies artificial baby food; but however convenient, practical and advanced this may be, when there is no compelling medical reason not to breastfeed, bottle feeding is an avoidable mismatch.

What is wrong with sleep?

During prehistoric times, people slept in two or more cycles a night, as the fire had to be watched. Several sleeps per night is a custom that was also common in our parts during the Middle Ages. People would sleep a few hours, be awake for a while, and then go to sleep again. This 'first nap' or 'dead sleep' was called *concubia nocte* in Latin, *prima sonno* in Italian and *premier sommeil* in French. Prayer books from the fifteenth century have been found with special prayers for saying in between the two sleep periods. There is also charming advice from a sixteenth century physician. He told his married patients to consummate their marriage betwixt the two sleeps, as not only would the enjoyment be greater during the nightly hour, but so would the chance of conception.

The upper crust slept in luxurious four-poster beds, above which a cloth had been suspended to ensure that no dust or vermin would drop into their sleeping open mouths. In many regions sleeping was done upright in a tight box bed, as people were afraid that too much blood would flow to their heads lying down in a flat position, and that this would cause brain haemorrhages (a good example of a cultural mismatch, because this correlation has never been proven).

It is likely that the eight-hour continuous sleep as we know it came into being fairly recently as a result of the industrial revolution. When, during the nineteenth century, labourers and other workers began to work for twelve to no less than fourteen hours a day (a mismatch with the work ethic of ancestral times, as we will see in Chapter 4), they were so worn-out that they slept through the night and no longer in two or more cycles. Medical advice was also to stop sleeping in two phases, especially for children. In a medical journal from 1829, parents were advised to severely reprimand children who, after their first round of sleep, turned over for a second nap.

During primeval times children and parents slept together, possibly under animal skins. Mothers had their babies near them and toddlers too slept close by. After the agricultural revolution this custom persisted. As accommodation became increasingly cramped, families in the Middle Ages slept together in one space, often alongside animals. No wonder they did not sleep through without a break: fleas must have driven them crazy.

With nineteenth- and twentieth-century prosperity came the

custom to give children their own rooms. For years doctors and professionals in mental health care recommended that children slept apart and should not be allowed to huddle with their parents. It was argued that children who slept in the parental bedroom had more sleeping problems and grew up into wimps.

This turns out to be far from true. Making children sleep alone compulsorily is another mistaken cultural innovation, a mismatch. Children who are given the option to spend the night in the parental bedroom – if they feel like it – from a young age, develop greater self-confidence, display fewer behavioural problems, are less susceptible to peer pressure, feel happier and are more content with their lives. And they are much less likely to show stress than children who do not sleep with their parents. As put by American paediatrician and educationalist William Sears: 'Over the past thirty years of observing co-sleeping families in our pediatric practice, we have noticed one medical benefit that stands out; these babies thrive. "Thriving" means not only getting bigger, but also growing to your full potential, emotionally, physically and intellectually.'

Adolescence!

When adolescents yet again blame their parents for everything, these parents can argue that it is not their, but the agricultural revolution's fault. Because of better education and health care the development of our body has fallen out of step with that of our brain. We are talking primarily about adolescence and about what is called the adolescent brain here. The hypothesis is that in primeval times our bodies and brains were well-balanced: as soon as someone was sexually mature physically they would be mentally as well. When youngsters became sexually active, they would be more than ready to have children. The agricultural revolution has caused our bodies to develop much more quickly, while our brains come trailing along behind. The young female body receives cues from its environment that instigate it to become sexually active, because it is fed with a calorific diet, for instance, or because the father is absent due to a divorce and the girl wants to bond herself quickly to 'a man about the house'. But mentally the girl is not ready to become pregnant (more about this in the next chapter).

The disparities between the development of the brain and body have obviously widened because we live in a social environment that is much more complex than that in primeval times. Boys and girls not only have an extended family, they have many friends, they acquire colleagues, they take part in a double life on social media that is as thriving as it is strangulating, they have to plan a career, earn money and possess all the cognitive and social skills necessary to become a successful adult. That's why physical development and brain development are out of sync. To put it in popular terms: our children listen less to their bodies and their bodies listen less to them. And they listen even less to their parents. Sulking is a mismatch.

Parenting missing cues

There are many challenges in the modern environment that hoodwink us when it comes to raising children. Firstly, we have to contend with the problem of large families, which did not exist in ancestral times and only came about after the agricultural revolution. Parental investment in their offspring diminishes considerably with their youngest children. Research shows that the youngest children in large families fare worse in terms of career, health and having children themselves when they are grown up.

Another societal problem that suggests a mismatch is the ever-changing and increasingly complex family make-up. Evolutionary psychologists Daly and Wilson have conducted an investigation into the so-called Cinderella effect: the increased chance of child abuse and neglect by step-parents and non-biological educators. In cases of child abuse and infanticide, step-parents appeared to play a bigger role than biological parents. Studies also show that parents who have both biological and step-children in one family, give preferential treatment to their own offspring over the other children.

Unlike in ancestral times, children nowadays are more likely to be surrounded by their contemporaries than by a cross section of several age groups. This leads to a great deal of competition, stress and struggle, which has an adverse effect on the learning process. Children learn things from older role models, in particular from elder children. In the modern system children spend the majority of their day with contemporaries, their competitors for now and later. Our thinking is that this makes the whole of society more competitive and individualistic. An attendant problem is that children are not able to practise their 'parental capacities' either (older children taking care of younger children in the group as role models, for instance), and consequently become worse parents than they might have been.

The educational system does not always stimulate in a way that is evolutionarily sound and in so doing invites mismatch. In our present school system the emphasis is on cognitive rather than social-emotional development. In other words: we would rather children learn information and compete with each other than them learning to live and work together. Nor is the infant brain always well-adapted to the information it has to process. Tests show that in the UK 20 per cent of children in primary schools have insufficient reading skills. All in all, 10 per cent of children are 'functionally illiterate'. It is also estimated that as many as two million British children have trouble with numbers and number work. In the US, thirty-two milion people cannot read. That's 14 percent of the popu-lation. Dyslexia and dyscalculia are the result of mismatch. The infant brain is adapted to an ancestral savannah environment in which writing and algebra did not exist. These are evolutionary novelties to which our brain is still trying to get used to.

ADHD

ADHD also suggests a mismatch. In 2000, it was estimated that 3 to 5 per cent of children showed symptoms of the disorder (extremely impulsive behaviour, problems concentrating, restlessness and learning difficulties); in later years this number rose to between 7 and 8 per cent, and if we wait another few years the time will come when everyone appears to have ADHD. In the media the condition is described as a fashionable disease. As a modern family you no longer count for anything if you do not have at least one child with a diagnosis of ADHD, whether it be self-proclaimed or not.

Research tells us that the disorder exists in all cultures, but is more prevalent in boys than in girls. Evolutionary scientists automatically look at what the adaptive function of this behaviour would have been in the ancestral environment. ADHD is characterised by impulsiveness, inquisitiveness and excessive exploratory behaviour. In some of our ancestors' environments it might have been advantageous to innovate: if you were reticent during times of scarcity you were less likely to survive it. In certain circumstances impulsiveness and inquisitiveness were the best traits to have in order to survive. This may have applied more to boys than to girls, who tended to stay closer to their camp. The question is whether in our society (in which we are not all that nomadic and adventurous) boys are worse off for this previously normal behaviour. ADHD might be a label we stick onto boys who have difficulty sitting still at school from nine to three thirty, with all the ensuing problems, conflicts and medication.

According to our mismatch theory there is no medication for the 'new' disease of ADHD. There are drugs that may lessen the symptoms, however. Danish research shows that the use of this ADHD medication has sky-rocketed between 2003 and 2010, mostly in the United States where 5 per cent of children of school age take drugs to temper their behaviour. It is estimated that this generates a turnover of between thirty-six and fifty-two million dollars. We suspect that children will continue to take medication for a quite a while longer yet.

You're getting older, dad

Our Palaeolithic ancestors faced many difficulties: they could get injured and infected, there were attacks from predators and many violent clashes with neighbouring tribes. Their immune system was working at full tilt, they were very mobile and their food contained fatty acids and carbohydrates in balanced proportions. The wild plants and fruits they ate (they were able to choose from hundreds of different species) contributed to an extremely well-developed immune system to ward off acute threats.

After the agricultural revolution our living circumstances changed at breakneck speed where nutrition, housing and social cohesion were concerned. In fact the changes happened so quickly that our immune system has not been able to keep up, even though on average we are getting ever older thanks to better hygiene, permanent availability of calorific food and increasingly advanced health care.

Our bodies are getting older, but our brains and immune system have trouble keeping pace. In ancestral times, when someone had reached a particular age, the brain gave off signals that their work was done. They could go and find a place in the forest to fade away into sweet oblivion. But in modern times our bodies remain intact, whilst our brains decline. Alzheimer's, dementia and cancer, too, are diseases that did not feature in prehistory – nor amongst present-day hunter-gatherers.

The question is whether our bodies are equipped to reach the age of a hundred or more. We are faced with deteriorating cell growth, badly functioning organs, defective body parts and a faltering immune system (no longer able to ward off infections) as well as a brain unable to manage any longer. Health services are running at full stretch to keep the elderly mobile with new hips, transplants, drugs and operations cheerily being proffered.

Enjoying your food

Experiment 1. When a child is presented with a bowl containing only blue M&Ms, they will eat just a few. And if they are given only green ones, the same thing will happen. But if they are offered a colourful mixture, then their consumption will shoot up. Confectionery is an exaggerated cue.

Experiment 2. When rats in a laboratory are given a fatty product, they will tuck in ravenously until they are sated – and then they stop. When those same rats are given something sweet, they will once again tuck in ravenously until they have ingested enough sweetness – and then they stop. It seems rats have a natural curb when they have eaten enough fat or sugar. But consider the situation when those self-same rats are presented with something that is both sweet and fatty. They tuck in greedily until they have ingested plenty by any measure – but they do not stop. The combination of sweetness and fat in large quantities is virtually non-existent in nature. So this is an example of a fake cue. Rats do not know when to stop when faced with such snacks and will stuff themselves. Just like humans, who do not know when to stop either. And it is humans who have come up with this food that is both fatty and sweet. A mismatch with gigantic implications.

In 2013 the commendable book *Salt, Sugar, Fat* by Pulitzer Prize winner Michael Moss appeared. It examines the way in which the food industry manages to persuade consumers to continue eating and buying demonstrably unhealthy sweet, salty and fatty products. One of the key terms in his argument is the so-called 'bliss point': a combination of salt, sugar and fat which is so ideal it stuns our brains and we carry on eating unthinkingly, even if we are no longer in the slightest bit hungry. Food scientists around the world are beavering away to establish this 'yearning point' for numerous products (soft drinks, savoury snacks, cereals, dairy, ice cream, ready meals, processed meat and fish). Everything in the product (from packaging, saliva-inducing additives to melting points and texture) has to contribute to the consumer's continued gorging. Who can open a bag of salty crisps and only eat half of it? What's more, the food industry's marketing budgets are dazzling: in America alone, 4.6 billion dollars was spent on marketing in 2012.

Moss' book is a shocking account of how the food industry has perfected both the bliss point and marketing, and in so doing has made entire generations fatter, more gluttonous and more unhealthy. The comparison with the tobacco industry thrusts itself upon us, because consumers have in fact been rendered addicted to stimulants with damaging consequences in the long run. Some food giants – fearful of legal claims like their counterparts in the tobacco industry – tried to reduce the amount of salt, sugar and fat, but the sales figures of these 'reduced' products dropped immediately. It goes without saying that these less unhealthy products disappeared off the shelves, for if there is one thing more important than ethical considerations it is maximising profit. Fortunately, due to pressures from governments and consumer organisations these healthy versions are now increasingly available again in supermarkets, showing that there are concerted attempts to deal with these food mismatches.

Travelling as a mismatch

We are travelling ever more frequently and to further destinations, with positive effects: worlds are opening up, our horizons are broadening and cultural and genetic in-breeding is disappearing. Yet associated with travelling are many mismatch problems, as our bodies and brains have not adapted to it all that well. The Dutch writer Cees Nooteboom once wrote that if you go to a far destination, your body travels by aeroplane but your mind chugs along by steamboat. Mismatch manifests itself in fear of flying, amongst other things. Primeval humans were not used to moving from one place to the next through the air, which is why many people hesitate to step into an iron bird. Then there is jetlag. These days we can get on a plane and twelve hours later arrive on the west coast of America. Israeli immunologists have studied the effects of jetlag on both mice and people. Long flights with a big time difference affect our gut microbes. One of the outcomes is that our bacteria are completely out of sorts for a period of up to two weeks, with an increased chance of glucose intolerance and even obesity as a result. People who frequently fly long-haul risk chronic health problems, according to this research.

Another mismatch resulting from travelling long distances is the spread of viruses and infectious diseases. Ever since the agricultural revolution, population density has increased spectacularly, and since the industrial revolution we have been moving about in ever larger numbers, especially since the advent of civil aviation. Increased global trade in poultry and cattle is also responsible for ever greater risks. Epidemic diseases such as influenza, Ebola, SARS, H1N1 (swine flu) and bird flu are potentially able to spread across the entire planet without any difficulty and have done so in the past. From history we know of some large-scale outbreaks of disease. The so-called Antonine Plague raged between the years 165 and 180 AD and is estimated to have killed five million people. Likewise the Black Death which wreaked havoc between 1347 and 1352 cost the lives of an estimated twenty-five million people, or a third of the European population at the time. The most likely front runner is the Spanish flu which caused between twenty and fifty million deaths between 1918 and 1920. AIDS, too, was and is a true killer, with an estimated forty-seven million victims to date.

During the times that the plague caused mayhem on the European continent, the way in which the disease spread was obvious: travellers carried it with them from village to village, from city to city. These days airports are the biggest distribution centres. Researchers from Northwestern University in Chicago developed a simulation method published in *Science* in order to be able to predict outbreaks, not measuring distances but by looking at how international airports are connected to each other. Present-day viruses simply travel by plane, just like us.

Viruses such as flu, Ebola and AIDS continue to adapt genetically. They take advantage of mismatch. Scientists are genetically modifying viruses that cause diseases, in order to control them and to find a cure. Other scientists are concerned about this, such as Stephen Hawking, who has spoken about the possibility of humanity being wiped out by a destructive genetically engineered virus, either through accident or design.

Health problems

Palaeontologists have found small bundles of medicinal plants in and amongst fossil remains which suggest that our ancestors engaged in medicine early on. They have also found prehistoric skulls with bore holes, drilled using flint tools. Skull trepanation is an ancient custom that was primarily used for insanity and madness. The fact that skulls have been found with carefully executed borings points to ancestral attempts at medicine.

Just like us, our ancestors obviously had to contend with illness and physical discomfort. We know that sick animals instinctively eat particular herbs when they have specific physical symptoms and we can safely assume that our ancestors did likewise. They will have known which of the hundreds of plants and herbs and (at least a thousand African) fruit species were poisonous and which offered relief for physical ailments. Health care will have consisted of observation and experimentation, until every tribe and nation knew which mineral, vegetable and animal materials had a supposed therapeutic effect or otherwise. This information will have been passed on from generation to generation and individuals who specialised in the knowledge of herbs, roots and fruits will have obtained a higher social status – and as a result may themselves have had better procreation chances. Becoming a medicine man or woman became a calling someone could aspire to. Herbs not only had a therapeutic effect, they could also be used to achieve a hallucinatory high, which was seen as a state in which contact could be made with the transcendental. Medicine men and women and the herbs they picked thus acquired a magical reputation.

Belief in the divine also played a big role following the agricultural revolution (and continues to do so in present times for those who attach value to spirituality, faith healing and forms of natural medicine such as homeopathy). And yet in several places on Earth treatment methods sprung up which were based on medicine as we now know it: involving clinical diagnoses and scientific knowledge of anatomy and syndromes. The oldest and best-known doctor was a man called Hesy-Ra, who worked as a physician for Pharaoh Djoser during the twenty-seventh century BC.

It was Islamic civilisation that was responsible for a big forward leap during the Middle Ages. Elaborating on ancient Greek and Indian medicine, Islamic physicians laid the foundation for the medical science of today. With the discovery of the microscope (in 1676, by Antoni van Leeuwenhoek) the link between diseases and micro-organisms, which until then it had not been possible to see, was laid bare.

Current medical science relies heavily on medication and the acute treatment of symptoms. Preventing illness plays a much smaller part, even if doctors would prefer that not to be the case. Nowadays, rare congenital genetic defects make up less than 5 per cent of all diseases. A large part of all the other complaints is rooted in our modern way of life, according to the Dutch evolutionary physician

Remko Kuipers. He estimates that around 90 per cent of Diabetes Type II cases, 80 per cent of cardiovascular diseases, and 70 per cent of strokes and bowel cancer could be prevented if, as a society, we pay attention to our nutrition, excess weight, inactivity and smoking. All consequences of mismatch.

Bad habits

On a café terrace, we overheard a conversation between two elderly men who were having a smoke outside the entrance.

'Everything okay down under?' one of them asked, inhaling wheezily.

The other one looked at him.

'Not for years.'

He lifted his fag and gestured despondently at it.

'These chaps every day for thirty years,' he said, exhaling smoke. 'I can't get it up any longer. They say it's a side-effect of smoking.'

'My God,' the other man said and looked at his own cig. 'Couldn't they've told us that a bit sooner?'

Whereupon he took a deep drag.

Smoking is the inhalation of smouldering, dried tobacco leaves. The question as to how old this custom is, is the subject of scientific debate. The Greek physician Hippocrates (c. 460–370 BC) believed the smoke of particular plant species was efficacious for some ailments. The original inhabitants of America had a long tradition of smoking, not only because tobacco, mushrooms and other plants had an hallucinatory effect in high doses, but also because of the idea that tobacco above all was 'a gift from the Creator', and could be used as a painkiller and a cure for flu.

The European explorers brought tobacco plants back Europe, whereupon the plant began its destructive victory march. In the Netherlands around 20,000 people die every year as a result of (passive) smoking, in America in excess of 480,000 and globally four million. Tobacco contains toxic substances such as tar, nicotine, carbon monoxide, nitrogen monoxide, arsenic, cyanide, ammoniac, acetic acid, polonium and many more besides. The list of diseases tobacco is (partly) responsible for is just as long: cardiovascular diseases, respiratory diseases, periodontitis, practically all forms of cancer, thyroid diseases, bone, joint and muscle diseases, back complaints, erection problems, deafness, macular degeneration, blindness, Alzheimer's disease and on it goes. In short, our body has manifestly not adapted to a smoking existence.

Nevertheless 28 per cent of the Dutch continue to smoke and this figure corresponds to the percentage of smokers in the UK (26 per cent). The Dutch TV-programme *De Rekenkamer* (*The Audit Office*) calculated in 2011 that smokers may cost society a great deal of money (around 2.4 billion euros) as they are ill more frequently and interrupt their work for a smoke, but they also fill the Treasury's coffers (around 2.3 billion euros). In the UK, a staggering 12 billion pounds is brought in by smokers each year in direct tax revenues. The tobacco industry earns approximately 230 million euros in the Netherlands; a favourable side effect is that smokers die on average four years earlier than non-smokers and therefore save 1.5 billion euros in public health, social care and pension costs. The cynic might say: keep lighting up!

Smoking is what biologists call a 'costly signal', what a peacock's tail is to a peacock. A peacock with crazy plumage on its behind exhibits: I am so strong genetically that I can indulge in these beautiful crazy feathers, so choose me as a partner! It's the same with smoking. Smokers, especially young ones, show off that they are tough and strong enough to allow all that poison into their bodies; these days they also flaunt that they have enough money to be able to invest in their costly hobby.

Unfortunately, research from the University of New Orleans shows that, on average, male cigarette smokers have a 41 per cent greater chance of erectile dysfunction than non-smoking men, which is not conducive to reproductive success. Men who smoke more than twenty cigarettes a day have a 65 per cent chance that they will not be able to get it up at a later age. And the bad news is: this chance is not lowered by stopping smoking. The message is simple: if you want to keep 'everything well down under', do not start in the first place.

Sunbeds

When our skin is exposed to ultraviolet rays, this not only ages the skin, it also increases the risk of various forms of skin cancer. Ultraviolet radiation is present in sunlight, but also in sunbed lamps. Research shows that for people who start using sunbeds before the age of thirty the risk of melanoma, the most serious form of skin cancer, increases by 75 per cent. UV rays can cause mutations in the skin's DNA and in so doing disrupt the cell's cycle.

Sunlight produces vitamin D, which in turn is good for our bones and immune system. Leaving aside the fact that an afternoon in the sun perks up our mood considerably, sunlight is literally of vital importance. The UVA light produced by most sunbeds does not stimulate the body to manufacture vitamin D. This is an example of a fake cue that cannot replace an absent cue (too little sunlight). There are lamps with UVB light that could produce vitamin D, but these contain far less UVB radiation than natural light on a sunny day.

The Dutch Society of Dermatologists and Venereologists (NVDV) called for a ban on private sunbeds. The number of patients diagnosed with skin cancer has tripled over the past twenty years and rises annually by 6 to 9 per cent. Tanning on a sunbed at home is said to be largely responsible for this. A mismatch, because we are seeking sunlight artificially to look healthier and more tanned and therefore more attractive – but it's only skin deep.

Depression

A depressed feeling (or depression) is a normal function of the brain, as a warning response to events that require attention. A stressful event like a divorce or the loss of a loved one can be the trigger for a prolonged mood disorder and a loss of vitality. It is likely that 'depressions' as we know them now occurred in a very modest form during prehistoric times. Like us, our ancestors will have felt the need to withdraw from a hectic existence to cope with a loss. This is also observed in present-day hunter-gatherers, even if their periods of social retreat do not last as long and do not feature in the present psychiatric classification system, *DSM-5*, the manual for psychiatric disorders published by the American Psychiatric Association.

Depressions become pathological when the deep despondency does not go away and is self-destructive. During the twentieth century doctors came to the conclusion that psychiatric disorders are located in the brain, whereby a distinction was made between internal causes (unfavourable circuits that may be genotypical or not) and external ones (the vicissitudes of life). A veritable depression industry has cropped up since then and some people believe depression has become the most common disease. Some research classes the Netherlands as the 'happiest country in the world', others claim we have well over a million people on antidepressants.

Research into mental illness amongst the Kaluli, a hunter-gatherer people of Papua New Guinea, shows that clinical depression is almost completely non-existent, despite the fact that these people, like westerners, are plagued by major setbacks such as illness and the death of loved ones. A hunter-gatherer life appears to be profoundly anti-depressant. In the words of clinical psychologist Stephen Ilardi: 'As they go about their daily life, they naturally wind up doing things that stop them from getting depressed, things that change the brains more powerfully than any medication. These range from exercising regularly and eating plenty of omega-3 fats to belonging to active social networks and getting enough sleep.'

We do not wish to suggest that depression can be easily solved or prevented by going for a run, eating some fish, maintaining friendships and having a restorative nap from time to time – but rather that there are forces in our modern environment which make dealing with stressful events more difficult. One of the most important predictors of depression, loneliness, did not feature within the strong social structures of our ancestors.

Our ancestors lived in Africa, a vast arena in which the rhythm of day and night was fairly constant. Having left Africa, people came to live in areas where in the summer the sun does not go down, or rarely does, and in the winter hardly appears, if at all. Medics talk of Seasonal Affective Disorder, or SAD (not to be confused with the SADD effect we will be discussing in Chapter 9), in people who become depressed when the days shorten. In high latitude regions (like Norway, Sweden and Canada) it is a fairly common phenomenon, while it does not occur in Iceland – probably because of the Icelanders' genetic predisposition. Symptoms of SAD are excessive tiredness, a disturbed eating pattern, irritability and general gloominess. In Japan as well as in Europe there are more suicides in the north than in the south, something SAD could be responsible for. Research suggests depression increases the risk of suicide fifteen- to twentyfold.

There are various treatments for the effects for SAD (artificial light and antidepressants), but the simplest is a temporary relocation to a region where the sun does shine in the winter. Why not book a holiday to ancestral environments a bit more often?

Suicide

Suicide is an evolutionary mystery. Is mismatch theory able to unravel it? According to an article in *The Lancet,* every forty seconds someone in the world takes his or her own life. Annually this amounts to a million people, 30 per cent of whom are in China. A cautious estimate suggests that, on top of that, at least twenty million suicides are attempted.

In 1897 the French sociologist Émile Durkheim published his standard work *Suicide*, a large-scale investigation into suicide. Although the book was heavily criticised later, many of Durkheim's findings led to a change in the prevailing attitudes about suicide. Durkheim noted that country farmers who move to the city had a greater risk of dying from suicide. He assumed this was because these farmers' social connections fell away rapidly. This accords with anthropological research into suicide amongst present-day hunter-gatherer nations conducted later (by sociologist Tony Waters, amongst others). Although violence was the order of the day in ancient times, suicide was an unknown and unthinkable phenomenon amongst hunter-gatherers leading a traditional life. This was until these tribes began to settle (semi-)permanently and modern life got a grip on them. Multiple studies show that groups that have experienced extremely rapid changes in lifestyle, score highly in suicide figures as a result of problems with alcohol, divorce, youth criminality, mental illness, illegal sexual behaviour (also called 'age-vertical sexual relations') and a total dismantling of social relations and society. It appears that many cues promoting social cohesion in a traditional community are lacking in a fast-changing society.

Durkheim's standard work also showed that more men commit suicide than women (although married childless women are an exception to this), more single than married people, more protestants than Catholics and Jews, more soldiers than civilians, more Scandinavians than other Europeans, and that more people kill themselves in peacetime than in war (example: when in 1866 war broke out between Austria and Italy, the number of suicides in both countries dropped by 14 per cent). That more men continue to commit suicide than women was also revealed in a VU University of Amsterdam study in 2014. In 2013, 1854 people in the Netherlands ended their own lives. This was a considerable increase on the previous years, when the number of annual suicides oscillated between just over 1300 and just over 1600. Clinical psychologists believe this rise was related to the economic crisis. The people who committed suicide were mostly middle-aged men. Men, according to professor Ad Kerkhof in an interview in the Dutch newspaper *Trouw*, are less deeply rooted in their lives than women: 'They derive their identity from one thing only: their job.' In primeval times there were no jobs and men were able to enhance their status in all manner of ways.

Why would anyone still able to pass on his genes successfully deprive himself of this prospect by dangling from a rope? From an evolutionary point of view this makes no sense. Young people who throw themselves from a tower block are no longer able to have children, and older people who leave life behind are no longer able to look after their children or grandchildren. Why is this destructive behaviour human at all?

Depressiveness is not specifically restricted to people, Frans de Waal showed. Negatives emotions go back a long way. Like us, chimpanzees for instance, are capable of joy and enthusiasm but also of boredom, grief and melancholy. Yet chimpanzees do not commit suicide – whereas we do.

A mismatch explanation, articulated by anthropologist Meredith Small, is that the primary function of suicide attempts is not suicide itself. The number of attempted suicides is much greater than the number of successful suicides. Small: 'Commonly called a cry for help, these acts do indeed change the life of the survivor as well as the people around them.' In the best case scenario a suicide attempt is seen as an alarm signal, whereupon it is up to the loving people surrounding the person who made the attempt to put things right.

Thus a suicide attempt becomes a life saver. In the worst case scenario someone actually dies. According to this theory, suicide would be an annoying side effect of a suicide attempt. As, because of all our cultural innovations, it is relatively easy to end your life these days – there are tall buildings and plenty of fast trains in our modern urban environment – this effect is much greater than in earlier times (more about suicide in Chapter 9).

Motorised mismatch

On 31 August 1869, the creditable northern Irish amateur painter and scientist Mary Ward took a turn with her husband and two cousins, pioneers in the field of steam technology. The vehicle in which they made this trip was a cart propelled by a steam engine. Steam cars had been around for a few years and a speed limit had even been introduced (four miles per hour in the country and two miles per hour in towns and cities). When the car took an unfortunate bend, the forty-two-year-old Mary Ward was thrown out of her seat and ended up under the wheels of the vehicle. She broke her neck and died instantly. Which made her the first motorised traffic fatality.

One hundred and forty years later traffic fatalities are a veritable global epidemic. Terrifying numbers of fatal accidents happen each year. In 2004, an estimated 1.2 million people were killed (including a quarter of a million children) and fifty million were injured (including ten million children). During the twentieth century there were around sixty million road deaths, at a rough estimate as many people as were killed during World War II. The possibility to drive from A to B in a car is a cultural innovation that is due to our desire to be able to cover increasingly large distances increasingly quickly. There are three factors behind road accidents: the driver, the vehicle and the road. We have been physically adapted to walk behind animals and gather nuts, but not to tear along in a machine on a motorway at a speed of seventy-five mph, alongside hundreds of other road users and in all possible weather conditions.

Much is being done, of course, to adjust aspects of this patent mismatch between our biological nature and technological developments (driving tests, examinations, maximum speed restictions, seat belts, alcohol limits, safety regulations, blind-spot mirrors, speed cameras, the creation of 'subjective experiencing of traffic safety' which stimulates road users to pay more attention), but until we ban the car there will be many more Mary Wards.

Modern bodies

In the same way that the human mind's creativity is ultimately the cause of all mismatches, there are hundreds, if not thousands of ways, approaches and ideas to cancel them out or turn them into a match.

Obviously, there is the option to do nothing at all. Doing nothing is always an option. Would it not be wonderful if in time natural selection of healthy-eating, non-smoking, highly active people were to occur, because no one is taking any more chances with a junk food-eating chain smoker? The latter's chance of having progeny has been reduced, after all. This would be rotten for them, but not for the generations coming after them. Tentative suggestions to deny hospital treatment to people with bad habits or to exclude them from insurance, soon smack of anti-social Darwinism.

A greater focus on better nutrition might expect more support. This starts with breasts. Breastfeed! The maternity ward of the VU Medical Centre in Amsterdam, for example, in 2015 introduced the practice that all babies staying at the hospital would be given breast milk – either from their mothers or from surrogate mothers (using that beautiful old-fashioned term: wet nurses).

We should also focus a great deal more on nutrition during our more advanced years. A lesson we could learn from our ancestors:

varied nutrition is better nutrition. And we do not mean a varied diet provided by the food industry, because as Christina Warinner has showed, processed food (be it chicken burgers, cake, cruesli or custard) is made for the most part with just three ingredients: corn, soy and wheat. We have adapted to eat as many different fresh ingredients as possible: ripe fruits hanging from a tree, waiting to be picked by us, or roots from plants that have just sprouted. Our ancestors ate fresh products and they demolished them in their entirety (not only the juice but also the flesh; not only the sugar, but also the actual beet). We can combat the entire nutrition mismatch by eating much less processed food, much less sugar, fat, and salt, and many more fresh seasonal products.

Activity. Repetition gets the message through. Thousands of studies tell us we do not move enough. Activity. Every day, we need to walk, cycle, run and exercise more. This is good for general health, good for holding off diseases like Alzheimer's and good for our immune system. The Romans knew it already: *mens sana in corpore sano*. Staying physically activity is essential.

Research shows that activity and nutrition go hand in hand. If you want to lose weight you will not succeed by just exercising. Ways to counter mismatch are: ban fried food in sports canteens, outlaw super-deals on meat, encourage cycle to work schemes, introduce car-free days, car-free inner cities, encourage working at standing desks, have meetings while walking instead of sitting around a table. The government could set up 'national fitness centres' of which all citizens would be free members (possibly run by deposed commercial fitness centres).

What about mismatch in education? In the Netherlands, there are now 220 Jenaplan Schools (named after the German town of Jena, where this teaching concept was initiated), where two consecutive years are placed in one classroom in the hope that the children will learn from each other. Group five children learn from group six pupils, who in turn benefit as well. Montessori Schools are based on a similar principle, but place three years in one group. This idea could be developed further, of course. Instead of eight year groups, schools could also create eight 'tribes'. Every year, each tribe receives a few new young children and loses a few older ones when they move on to secondary education.

Modern humans descend from hunter-gatherer cultures in which there was a clear distinction between male and female roles, which may explain the average differences in spatial and verbal qualities. Let's investigate how we can make boys less boisterous, less competitive and more empathetic, and how we can organise technical education in such a way that girls do even better and enjoy it even more. But it is irresponsible to dismiss the differences that are there – from a scientific as a well as from a social standpoint. If we want to do something about prostate or breast cancer, everyone thinks it goes without saying that we should focus on one single gender. But when it concerns demonstrable, evolved gender differences in 'talents, temperaments, traits and tails' we are often blind to it. This may be the reason why boys and girls do not always have an easy time in our society.

Let schools put more emphasis on the development of social behaviour, so that children are better prepared for our socially complex society. Reading to children in all grades can be crucial. We have (as we will see in Chapter 9) a narrative brain. The importance of telling fairy tales and stories cannot be overestimated. Many articles circulate online that extol the benefits of reading and telling stories, but according to evolutionary psychologist Peter Gray one reason really stands out: 'Stories provide a simplified simulation world that helps us to make sense of and learn to navigate our complex real world. The aspects of our real world that are usually most challenging, most crucial for us to understand, are social aspects. Knowing how to deal with evil as well as love, how to recognise others' desires and needs, how to behave towards others so as to retain their friendship, and how to earn the respect of society at large are among the most important skills we must all develop for a satisfying life. Stories that we like, and our children like, are all about that. They are not *explicitly* about how to navigate the social world, in the way that a lecture might be. Rather, they are *implicitly* about it, so the listener or reader has to construct the lessons for themselves, each in his or her own way. Constructed lessons are far more powerful than those that are imparted explicitly.'

Another hint: children do not benefit from too many restrictions. Give them more responsibility. Compared to other countries, the Netherlands has a tolerant policy towards sex and drugs. Yet despite our liberal dealings with sexuality the age at which young people lose their virginity is by no means lower than in other countries in Europe and the world. Or perhaps it is because of this tolerant policy. Our problems with drink (and drugs) are a lot more manageable then in countries with very strict anti-alcohol legislation such as Sweden, for instance.

The role of the extended family in parenting can be underlined by making it financially and fiscally easier for families to have grandparents live with them or nearby. Family leave – normal in Scandinavian countries – should be encouraged more. Large families should get more help and supervision. Younger children are less likely to receive full parental care than their elder brothers and sisters. A British study showed that the youngest children in large families receive less nutrition and are vaccinated less against all sorts of infectious diseases. They are also smaller and on average have a lower life expectancy. Cherish your youngest children, in other words.

To conclude, we urge you to take your (newborn) babies outside much more often than is common these days. Our bodies are equipped for a life in the open air. There is evidence that people who spend a lot of time out of doors are much healthier than those who do not. By walking in the open air we invite more bacteria onto our skin, we manufacture more vitamin D and our immune system works better. As part of a British study, babies were taken to a children's farm shortly after birth, and were taken for frequent walks in the forest. This benefited their general health. The motto is therefore (not only for babies): go outside, roll in the mud, walk, be active, play. It can be as simple as that.

CHAPTER 3

A crazy little thing called love

Your life started when two cytoplasmic genes fused together. Since then you have developed into a breathing pouch containing hundreds of billions of cells capable of holding a book. How these two initial cells came to be together is the story of your life, of life itself, the story that is being retold day after day and that Freddie Mercury sang of as 'A Crazy Little Thing': love. At this moment in time love is being celebrated in thousands of cafés, at parties, during concerts, at work, in the supermarket over the freezer compartment, in shopping centres, on street corners, in church choirs, canteens and clubs, and even online in chatrooms and on dating sites. It does not matter where the story is set, while you are reading this two people will meet each other somewhere on the planet, fall in love and eventually let their cytoplasmic genes fuse together. We cheer them on.

Let's assume that our story takes place in an entertainment venue in London, Lancaster, Nottingham or Newquay, or any which other place where humans live. We see a queue of partygoers waiting at the entrance. A broad-shouldered bouncer and a high-heeled 'door bitch' separate the wheat from the chaff. No courtship display for individuals who are too young, too old, too scruffy or not bang on fashion. During the middle of the 2010s a club is obviously not the only place where people meet potential partners, but it is one which gets the imaginative juices flowing. Using an elegant sociological term we can safely call it a 'mating market'. We pan from the cloakroom to the main hall. People are well-turned-out, the men wearing tight shirts and trousers and the extravagantly made-up women dressed in clothes that leave little to the imagination. Behind the long bar, bartenders are shaking complicated looking cocktails with names like Sex on the Beach and Adam & Eve; alcohol is the pre-eminent social lubricant. The different ethnic groups are mingling amicably, visitors come from far and wide. The VIP-area, where the better-heeled, famous, expensive champagne swigging guests may strike up a conversation, is strategically positioned. What is striking is that men make women laugh more than vice versa.

We take a peep at the Ladies, where clusters of women are touching up their make-up. Suspended from the wall are chewing gum, condom and tampon machines. They gossip and discuss the men. One woman pops a pill, another one The Pill, as she had forgotten to take it that morning. *It's ladies' night and the feeling's right.* Back on the dance floor we register a pumping beat and strobe lighting that tries to send the excited crowd into a trance. The clubbers send each other innumerable signals. Without them being aware of it, everything this evening – or for scientific accuracy, let's says a great deal – is dominated by the search, sublimated or not, for sex.

A large study on going out (*Het Grote Uitgaansonderzoek*) conducted in the Netherlands by the Trimbos Institute in 2013 shows that 22 per cent of men go out with their regular girlfriend and 38 per cent of women with their regular boyfriend. Those without a partner are largely on the lookout, a British study confirms: 80 to 90 per cent of the singles visiting clubs and pubs would not mind stumbling upon a partner – or in more vulgar parlance, to score that night. Only a small percentage actually do: courting and conquering someone's heart is a rather precarious business. There are many more hurdles to jump before our two clubbers end up in bed in order to allow their genetic material to fuse together.

Research into Cupid's arrows reveal that two thirds of people believe 'love can strike at first sight', but that they do not expect it to lead to much. It also appears that not many women have this romantic notion, but mostly men. 'At first sight' is an elastic concept. It would be better to speak of 'love within an hour'. Some chatting, prying and scheming has to take place before love strikes. Meanwhile the couple judge each other on various traits such as similarities, humour, dependability and intelligence. Subconsciously, she focuses more on his eloquence, from which she derives his intelligence, social status and income, and whether he will be able to make her laugh. He is a little more interested in her looks, youthfulness and whether she laughs at his jokes. Men fall in love more easily in any case, call themselves more romantic and want to go to bed with a woman sooner than vice versa. There is an easy biological explanation for this. In women, the number of follicles (baby eggs, as it were) dwindles with each menstruation by a thousand or so, until around the age of forty-five when there are none left, whereas men have a nigh-on inexhaustible availability of their genetic material up 'til their final days. What's more, the burdens of pregnancy are unequally distributed between the genders. While she is saddled with a nine-month pregnancy, followed by a long period of breastfeeding, one seminal discharge in the right place can bring him an identical reproductive advantage.

So men are mad-keen on going to bed with women. In 2006, an investigation was conducted on how keen. Scientists asked attractive young women to start chatting to male students and to ask them fairly early on whether they fancied having sex. Three-quarters of the students did not have to think long about this. Yes, of course. Then the experiment was repeated with attractive men speaking to female students and asking them if they wanted to go to bed with them. One hundred per cent of women did not have to think long about this either. No, of course not. Literally zero per cent of the women said yes. The outcome of a fundamental difference between the sexes in their evolutionary interests.

After a sparkling night of dancing, talking, laughing and showing off we see the club gradually emptying out. The odd couple has got together, thanks to alcohol or not. One specific couple, who did not know each other until a few hours ago, are extremely charming. They have flirted and danced, she has checked out his pumped-up torso and he her Wonderbra. Her make-up makes her seem just that little bit more attractive than she probably is. They kiss and we avert our eyes for a moment. He is instantly in love, she keeps her distance. Arm in arm, they sway out of the club in a sweet glow. We cast one last glance at our Romeo and Juliet. Love, this crazy little thing.

The most monogamous ape

Is love the answer? Is love in the air? Is love one soul in two bodies? A battle? A force that can turn foes into friends? Is love rating someone else's happiness more important than your own? Is love a fruit anyone can pick? Is love doing the dishes together? Not letting his dirty socks get to you? Google 'love is . . . ' and you will get more than 200 million results.

Love is, in scientific terms, a psychological adaptation. Everything in nature is geared towards more reproductive success: traits with a reproductive advantage outstrip traits without. Evolution has seen to it that men and women find one another more attractive when sexual partners have genes that complement each other.

Humans need a great deal of parental care until quite an advanced age – especially compared to other primates – in order then to be able to have children themselves. In evolution, if a particular trait produces reproductive advantage, this trait will reign supreme over traits that do not have this advantage. That is how these traits spread through a population. Children of parents who do not get involved in their upbringing have a demonstrably lower chance of survival than children of parents who actively take care of their offspring. In other words: lengthy parenting produces evolutionary advantages.

The obvious question is: 'Who looks after the children?' The biological solution is that there is often just one parent who takes care of the young, usually the mother. Males can secure a female by bringing her food: 'If you bear my children, I'll make sure there is enough food.' This courting ritual, courtship feeding, can be found amongst numerous animal species.

Various ape species share food actively with each other within their group (for example the meat of a doe). Some share more generously than others, but not without a hidden agenda. Male chimpanzees and macaques have been shown to give meat to ovulating females, with the idea that they will be able to copulate with them in return for this. This is called the 'meat-for-sex-theory' in scientific speak. 'Meat-for-copulation' is a form of future planning. For them.

Human children, in contrast to the young of other primates, have to be protected and fed for such a long time that it became necessary for both parents to take on this role. An interesting challenge in our evolution was how fathers could be induced to invest in their offspring once the deed had been accomplished. This is no child's play. For how does a man know for sure that the genes to which he sacrifices his savings and free afternoons are his? DNA-tests have solved this most delicate issue of paternal uncertainty, but who will take such a test voluntarily?

Solutions for evolutionary problems are called 'adaptations'. The adaptation to resolve the problem of paternity was the introduction of exclusive couples or monogamy. This was advantageous for the male, but also for the female. The ancestral male would have greater assurance that he was caring for his own offspring and the ancestral female scooped a 'hired gun', a muscleman who could buckle down to protect her and her children against other men with bad intentions. In order to make pairing off permanent, falling in love was introduced, a feeling that makes our hearts beat faster when we are near our partner (but this is too romantic an image; in fact, our brains produce quantities of drugs that would cost a street dealer fortunes). This feeling is followed by love, another evolutionary adaptation. Love is an insurance policy for the future. It stops men from looking for another woman as soon as a child has been conceived and women, for their part, give men the assurance that their children are also his children. Our species ended up monogamous, or rather, monogamish. We are the most monogamous ape, that's one title we have managed to bag as a species!

Evolutionary scientists have three key questions when they look at a particular trait: how, when, and why. The question as to how love evolved in people is interesting, but of no particular significance to our argument. Maternal bonding, in other words the bonding between mother and child (a feeling that ensures that the mother will look after and protect her offspring) employs neurological processes and a host of hormones such as the 'love' hormone oxytocin. The love between two partners probably employs these same processes and neurotransmitters. When bonding between men and women became important for human reproductive success, there was no need to think up a new mechanism, because a comparable process already existed: the love of a mother for her child and vice versa.

At which point in the evolution of humanity did love come on the scene? The answer to this is vague, for obvious reasons, because love does not fossilise. There is so much bickering about this question in science that it almost leads to bloodshed. If you think you have the answer, you won't have heard the last of it. We could go into every theory here, but the simple fact is, (monogamous) love appeared.

Whatever the case may be, this was sometime after our ancestors had split into us and the chimpanzees (around five to seven million years ago).

Chimpanzees have no truck with monogamy, so why should we? Monogamy probably became important when humans moved onto the savannah and started to walk upright. This resulted in our pelvis narrowing. Surmountable for men, but women had to contend with a narrowed birth canal. This was less of a problem as children were born more prematurely. Compared to other primates and mammals, human females drop endearingly helpless tots. Whereas lambs and calves are able to walk and jump an hour after birth, it takes human children roughly eighteen months to get to this stage. At birth, children have a mere five functions: crying, drinking, sleeping, growing and emptying their bowels and bladders. That is a hundred or so functions too few to look after themselves (these days they also have to be able to cook, drive a car and use a cashpoint). As children were being born increasingly helpless, joint parental investment grew in importance. Love between partners makes this joint investment easier, a love that ensured fathers did not leave their lover and children to fend for themselves following birth. The genes of the fathers who did abandon their wives and offspring gradually disappeared from the gene pool. Love became a survival mechanism, 'the survival of the dearest'.

Love and monogamy enable men and women to invest together in the children requiring care until they have learned enough to live independently. This is eighteen years down the line, however (it was a little sooner in ancestral times). The question is why do human children have to learn so much. This is partly due to the fact that our battle for survival is more complicated than that of chimpanzees, for instance, that are able to survive independently from the age of three. Human children had to know a great deal more: how to hunt in a group, how to survive in different climates, how to hold out in a complex social group – and they needed to know about culture. In order to end up as a cultivated ape, our children had to mature increasingly late and our ancestors made greater and longer investments. This was partly why it made more sense to have one regular partner. Monogamy was not a luxury, but a sheer necessity.

Having your cake

Monogamy is just one side of the story. It is the compromise in the battle between the different genetic interests of men and women. Generally speaking, a man benefits from having several children, preferably with several women, because sperm is cheap and available in far larger quantities than the females' eggs. Alongside monogamous systems there were also polygamous ways of living (which can still be found in parts of Africa, the Middle East and strictly orthodox American communities, for instance). Some men have the means to support a second and sometimes a third family in addition to their primary wife and children. Providing for a second wife is a big investment for a man, because during humanity's two million year-long history, scarcity tended to be the defining theme. Only exceptional hunters were able to have several families. Amongst the Yanomami, a warrior people in the Amazon rainforest, we see that only the best and most deadly hunters and warriors can afford an extra family. During prehistory there were presumably also migrant men who supported families in several camps, but that was more likely to be an exception rather than the rule.

But the female has contradictory evolutionary interests as well. Monogamy may be a good strategy for her, but is it also the smartest choice for obtaining the best gene material? In a polygamous system it can be beneficial for a woman to enter into a relationship with a man who has several women already; better to be a prosperous man's second wife than a pauper's only one. Another way to acquire good genes is to be unfaithful with a genetically better equipped man than your own partner, because the latter cannot know for sure that he is the biological father anyway. This is another area in which chimpanzees have one over us. Female chimpanzees mate with a large number of different males, so that each male can breezily think he is the father. And this is advantageous, as the tendency towards infanticide hardly exists amongst chimpanzees (as opposed to amongst gorillas, for instance).

An indication that we people have a polygamous streak can possibly be deduced from our males' balls. Male chimpanzees have extremely large testicles, male gorillas very small ones; male human ones hang somewhere in between. It is thought chimpanzees pursue the battle for the eggs of a female as far as her vagina by inseminating her with huge loads of aggressive sperm cells to kill other males' semen, under the slogan 'may the best spermatozoon win'. In a world without monogamy, kamikaze sperm would not be an implausible adaptation, but the question is whether the theory might also apply to the predominantly monogamous ape species *Homo sapiens*. The balls of our distant cousin the gorilla are tiny, which fits in with his reproductive strategy. A gorilla does not need large balls, because the only thing he needs to do is to ensure he becomes dominant in his group at which point he can claim all females. The alpha male amongst gorillas has the exclusive right to a harem. In terms of ball size, humans sit halfway between chimpanzees and gorillas. This leads to the supposition that we may be monogamous, whilst at the same time have an inclination to compete sexually for several partners. The presence of kamikaze sperm, as in chimpanzees, has not been found in our species, which leads to the supposition that women do not allow themselves to be impregnated by several men (group sex is still a rare phenomenon).

This 'having your cake and eating it' position does not alter the fact that there are individual differences in sexual behaviour amongst humans. A large proportion of people, perhaps as much as 40 per cent, have once or twice played fast and loose with our ingrained monogamous way of life. It has been estimated that a number of children (4 per cent, with peaks of no less than 30 per cent in some cultures, according to a coordinated international study), are the result of 'paternity fraud': they are not the biological child of the man they call their father. And that leaves us on the monogamous side. But as soon as a male has the means, because he has acquired status or amassed a lot of 'meat', it is not unthinkable that he will start looking for some extramarital sex opportunities, thus increasing the chance of his booting a child into the next generation. A female may also desire 'sex on the side', which may lead to her encountering someone with a better genetic profile than her present partner, who will – unbeknownst to him – raise a cuckoo in the nest.

Studies also show that there are people who prefer short-term relationships and people who seek long-term relationships, and that these are two genuinely different personality types living side by side. Both strategies can be found throughout the planet and this will have been no different in ancestral times. Maintaining a relationship alongside your official one is no picnic, even though it will have occurred amongst both men and women. Some anthropologists believe that when men were out hunting, some women in the tribe yielded to the attention of other men ('special friends', as the American biologists Helen E. Fisher calls them, would lure women with presents in exchange for physical favours). Nor would some men in the group have been all that fastidious about their choice of lover. But it is likely that social pressure prevented too many excesses. The groups in which extended families wandered over the plains were small and easy to keep track of, and public disgrace following adultery was often murderously great, especially for women. 'Slut phobia' is evolutionarily deeply ingrained in women.

There are other reproductive strategies in the animal kingdom which may be applicable to humans. According to game theory, when the majority of people raise their children predominantly in a monogamous context, there will be room for some uncommon sexual strategies that also lead to reproduction. These are niche strategies. The 'sneaky fucker' strategy is an example, a strategy that is geared towards men attempting to impregnate women when their partner is absent. Research amongst orang-utans shows there are two types of males. The first is big and dominant and has his own territory to entice females. The second type of male is smaller and does not have his own territory. As they resemble females in stature they are ignored by the dominant males, allowing them free access to the females that cross their path.

By way of illustration: in the human world some women have a so-called 'chat friend', a man to whom they can pour out their hearts without there necessarily being a love relationship or sexual attraction. At any rate from her perspective. In an unguarded moment, a woman may throw herself onto the friend with the listening ear (to his delight, because this was what he wanted all along). Another getting-off trick is the so-called 'posing as gay strategy', whereby a man puts on a feminine front or even pretends to be gay. In *The Feast of Love* (*Het feest der liefde*), a collection of stories by one of us from 1995, a senior student explains to a first-year student: 'Do you want to know what helps when trying to get off? Being gay. Being gay and saying that you've never done it with a girl, but that 'you should maybe, just maybe' give it go. Brave girl who will not jump on top of you then and there.'

Another niche reproductive strategy is one we can only refer to as rape. In *A Natural History of Rape* biologist Randy Thornhill and anthropologist Craig Palmer explain that amongst many animal species, rape is a fairly common form of reproduction. Drakes do it continually, as everyone who looks out onto a duck pond will confirm with discomfort. What is the situation for humans? Is rape a deeply-rooted strategy for men to boost their reproductive success or is it a modern cultural phenomenon? There is an assumption that particular kinds of men who, because of their appearance or social position, are not able to enter into a long-term relationship with a women, will nevertheless try to spread their genes through rape. In prehistory this presumably was an extremely rare and dangerous reproductive strategy, because social networks were so tight that a rapist ran a major risk of being banned or worse.

Suppose one in a hundred children was born as a result of rape in early hunter-gatherer communities, it would mean that this strategy would have been passed down to an extremely limited extent. The genetic advantage of this strategy will increase when the chance of sanctions decreases (more about this later in the book). In present times we see that our manners and networks are much less tightly-knit, which reduces the chance of exposure for rapists.

Homosexuality

What about homosexuality as alternative sexual strategy? From a Darwinian viewpoint exclusive homosexuality is puzzling as it tends not to lead to offspring. Yet, at the same time research suggests that homosexuality is more than simply a cultural lifestyle choice. Studies conducted across various cultures show that about 5–8 per cent of the population has an exclusive homosexual orientation. Among them are twice as many men as women. There is evidence from twin studies that homosexuality 'runs in families'. Biologists have discovered homosexuality in many other animals, from baboons to dolphins and from penguins to worms. There are various adaptive explanations for homosexuality, one being that by not investing in one's own offspring an individual could help raise the children of brothers and sisters. Another untested theory argues that groups with homosexuals run more smoothly as there will be fewer conflicts over potential mates. A third explanation that has been offered suggests that women may be attracted to gay men, because they are more empathic and cooperative. Partnering a gay man may have ensured better care for children. Although this does not explain exclusive homosexuality, it is an interesting possibility. Could homosexuality be a mismatch? To show this one needs to find data that under certain cultural conditions the percentage of homosexuals in a population goes up. This could be because gay men and women hold a relatively high status in society (e.g, as artists, designers and sport stars) or because there are societal norms favouring a homosexual lifestyle such as equal rights to marry and adopt children. At the moment we have no evidence to suggest that homosexuality, or LGBT-orientations more generally, are the result of a mismatch between human nature and particular cultural tendencies. Thus, for the remainder we focus on explaining heterosexual love from a mismatch perspective.

Love in ancestral times

Let's look at what love looked like in ancestral times. We will travel back to the African savannah, somewhere near Tanzania, in about mid-100,000 BC. We see a number of clans from all corners: broad-shouldered hunters, women doing most of the talking, young, old, sloppy and gracious, all in one big mix.

We skim the savannah where we see a large group of ancestors sitting around the campfire. Everyone is beautifully decked out and the teenagers especially have been decorated with the finest ochre patterns. The different age groups, partly separated by gender, are sitting together. The men drink an alcoholic brew and discuss war and peace. Strategically at the centre, the elder leaders of the tribe are conferring with each other.

The women, too, are sitting together. They are discussing social relations and attractive fellas. 'It's Stone Age night and the feeling's right.' Around the campfire meat is roasted, something is simmering on the fire; there is laughter, singing, dancing, and chatting. Signals dart back and forth: the people present will not be aware of every single thing, but everything – or for palaeontological accuracy, let's say a great deal – is dominated by arranging offspring.

Would 'this crazy little thing called love' really be a topic of conversation, or did the concept not yet exist in ancestral times? For an answer to this question we can go and have a look at cultures which have been oblivious to the agricultural revolution. The many so-called 'traditional societies' of past centuries have been examined and charted by western anthropologists, explorers and adventurers. They studied Australia, for instance, where original hunter-gatherers lived unhindered by White Man until the eighteenth century. The aboriginals were not able to develop agriculture – even if they had wanted to – because of the dry climate. They lived like our ancestors who wandered around the savannah for two million years. The inhabitants of Australia were still in prehistory when the country was opened up by Europeans: they did not beat metals, had no script to write in and did not make pottery. They hunted and gathered.

To the original inhabitants the arrival of the Europeans was the beginning of the end, but the Australians were a gift to ethnographers and other scientists. There were numerous tribes, clans and ethnic groups in Australia, and many different languages were spoken (some say no less than 250). Moreover, there were probably many cultures that must have resembled the societies that created us.

One of the largest tribes in Australia was called the Aranda, a people that lived in the wide expanse surrounding the centre of Australia, Alice Springs. This nation comprised a few thousand people, subdivided into groups operating independently. These groups would gather once in a while, but usually the small clans, consisting of a few families, moved around in their territory, exactly as our ancestors in Africa did for millennia. By our standards the Aranda lived simply: they chased marsupials, dug up lizards, ground wild plants and in the evening they delighted each other with dance, songs and stories around the fire. They did not own property; there was not much social cohesion and leadership was earned through respect for someone's capacities as hunter, warrior or medicine man.

The only thing that was truly complicated amongst the Aranda, much more complicated than in our present society, was this crazy little thing called love. The Aranda matrimonial rules were complex, strict and binding. No love at first sight, or at any rate it was strictly forbidden. In small, closed communities there is always the danger of inbreeding due to the fusion of recessive genes which can lead to all kinds of diseases and even infertility. These problems could be averted if people mated with the right partners.

The Aranda developed a complex system of 'kinship system groups' and 'sections' and 'subsections'. The son of a man from a particular kinship system group would end up in a different section from that of his father. In *A Brief History of the Human Race*, historian Michael Cook described the process: an Aranda male from the north of the territory was obliged to marry a female from the corresponding section of another kinship system group from the south. At times the tribes would get together for tribal conventions during which partners would be allocated and marriages conducted. Not all Australian hunter-gatherers used these complex marriage rules, but they all had a form of regulation. This shows that freedom of choice in sexual partners is not automatic in the evolution of humans. Relationships were often entered into with the involvement of family and based on considerations that had little to do with this crazy little thing called love.

Back to the Tanzanian camp fire, a hundred thousand years ago. After an exciting evening of dancing, talking and laughter, a few boys and girls are allowed to get to know each other better. One specific ancestral couple, who have known each other from previous conventions, have been given permission to get married. There are no kinship rules to stand in the way. At the end of the festivities we see the couple sneak off quietly into the forest. The boy demonstrates he is a good hunter by giving her a small piece of meat he had stashed away for her. She has tasted the meat, felt his muscles and had a furtive romp with him. They leave the camp in a sweet whirl. Whether they will go as far as fusing their genes together that night is something we no longer witness, because our ancestors, too, have a right to privacy, especially as – should they go too far – they would do so at a discreet distance far away in the forest or in a cave, as befits members of a monogamous species. We cast one last tender glance at our primeval Romeo and Juliet. Without them, we might not be here.

For ever and ever

Love Conquers All, the Deep Purple song goes, and many of love's mechanisms have passed the tests with flying colours. The feast on the savannah has been replaced by the disco and the piece of meat by an expensive cocktail, but in essence the boy-meets-girl story has been the same for centuries. Our primary senses – sight, hearing, smell and touch – are *Von Kopf bis Fuß auf Liebe eingestellt* ('from tip to toe ready for love'). We have thousands of ways, both conscious and subconscious, to send signals to potential partners, from smells we waft about to the movement of our limbs and speech – all evolutionary inheritances.

In ornithology, the area where male and female birds come together in large numbers to suss each other out and compare each other is called a 'lek'. The air is full of primordial signals for selection. The German ethologist Irenäus Eibl-Eibesfeldt has done a great deal of careful research into flirting behaviour. During the 1960s he developed a method to study test subjects without them noticing, to see how they observed each other. From Samoa to Papua New Guinea and from France to Japan he filmed 'flirting girls', which led him to the conclusion that flirting, throughout the planet, happens according to a fixed pattern of smiles, eyebrow lifting, pupil enlargement, head and eye movements, staring, bending, twisting, etc. According to Eibl-Eibesfeldt this behaviour is not culturally acquired but determined by evolution and – before humans became humans – it had developed to signify sexual interest. Opossums, horses, tortoises, albatrosses and countless other animal species do more or less the same, so no animal behaviour is alien to us.

Love may be blind, but it is top of the class in the smelling-stakes. Nature makes sure that we look for partners with complementary immune systems, so that our offspring have maximum protection against inbreeding and communicable diseases. Love is . . . the best of both bacterial resistances for the baby. Smell is an important indicator and therefore largely determines choice of partner. So-called MHC genes, which play a role in the immune system, set in motion a complex interaction between skin bacteria and odour recognition. Research shows that women prefer men whose MHC genes are different to their own. These are the men whose smell they like. If a woman does not like a man's smell, or even thinks it stinks, it can suggest they both have very comparable MHC genes and immune systems.

Through our physique, hairstyle and choice of clothing we display our social status, and our dance moves are literally a showcase. According to Darwin's theory of sexual selection, dancing is a prime way of showing you have a symmetric body, in other words have good genes. Try and dance gracefully when one of your legs is a few centimetres shorter than the other. Equally well known is this: male club-goers puff up their chests, which makes them seem broader than they really are (walrus males do the same before they have a scuffle over who gets the female) and women walk bolt upright on high heels, accentuating their secondary sex organs – breasts and buttocks. Their parading means something like: 'I may be available and have good genes'. The high heels denote: 'I can't run away so protect me!'

If all the non-verbal contact through eyes and gestures goes well, it is time to get acquainted. Even then a lot can go wrong, because people may have attractive bodies and faces, but social position, ambition, language and education are also important in the choice of whether to go home with someone or not. Men, though, generally feel much less strongly about these things than women. (Or have we already said this?)

Right across the millennia, an important icebreaker in direct contact has been humour. There is a direct line between sexual attractiveness and the extent to which men and women make each other laugh. Evolutionary psychologist Geoffrey Miller compares humour to a peacock's tail. You can use it to display verbal qualities, intelligence and creativity to the opposite sex. It is moreover a 'costly signal' (as we said, just like a peacock's tail), because not everyone has a sense of humour and it is therefore quite difficult to feign.

An American study by psychologist David Buss and colleagues revealed that women find GSOH (Good Sense of Humour) extremely important in two different relationships: in men they want for brief sex and in men they would like to marry. Many studies reveal that wit is the most effective weapon for men who want to seduce a woman. Naturally women also fall for things like appearance, status, money and odour, but an amusing man is very much one up on his humourless counterparts. Laughter is the way to a woman's heart. People have to get along for the rest of their lives, and humour can be a powerful weapon to get through the dull winter evenings in front of the TV. Incidentally, which humour tap is being pulled is of importance. Our own research shows that women appreciate jokes at the expense of other people (like jokes about the Irish) far less than self-mockery and other self-deprecating humour, except when they are ovulating. Then they rate men with a dominant style of humour who put down their competitors.

In *Why Women Have Sex*, by psychologists Cindy Meston and

David Buss, 1006 women gave no less than 237 different reasons why they went to bed with men (one woman was bored, another felt the need to boost her self-image and yet another was revenging herself on her unfaithful partner). The research showed that there are two traits women find sexually extremely attractive: humour and self-confidence. Which is good news for men who cannot bring into play a broad chest and yet do not lack self-confidence over this. You could even argue that the large number of stand-up comedians who are 'physically challenged' suggests that humour is an alternative mating strategy for men who would not get a look-in on account of their appearance or status (this may also apply to writers and scientists, but we will not go into that just now).

In order to strengthen the bond between partners, solemn ceremonies related to courting have been introduced since the agricultural revolution. Infatuation morphs into love and this requires a vow. However powerful the feeling and the accompanying emotions may be, unarguably there comes a moment when people in love desire more than some marginal noncommittal dating. In the end we want a commitment. When entering a love relationship there is always the danger you will be betrayed. The person you are in love with may behave as if he or she is also in love and call you 'the most beautiful person in the whole wide world', but we all know that he or she could whisper this to someone else just as easily. Love is ... sharing each other's lives, emotions and property. You cannot yield to another until you trust each other and this is something you demonstrate in a public ritual. A way to promise love is by giving each other objects: not cheap rubbish, nor things that can easily be passed on to someone else. This is how the giving of rings came into being. People invest in an expensive present that is only meaningful in an exclusive relationship. Rings are a way to make sure a partner is not soft-soaping you and is prepared to invest in your offspring. But: people are so affluent these days that even an expensive wedding ring no longer offers a guarantee for staying together. If you want to leave your partner you can just take your ring to the pawnshop!

Whereas the rituals are public, the consummation of love is principally a private affair. Public sexual behaviour is one of the strongest and most widespread taboos there is, according to anthropologists. While some liberal societies now accept lovers kissing each other in the company of others, pretty much everyone seeks the seclusion of a bedroom, broom cupboard, car or hotel for the actual sexual act.

Some evolutionary scientists believe we do this because we are defenceless against attacks from predators and enemies during sex. In moments of lust, it would not be such a good idea to lie in the middle of an open field. Other researchers presume that our predilection for seclusion during sex has to do with a fear of jealousy and aggression from third parties. Bystanders might get the idea into their head to have a go at impregnation as well, and we like sex to be an exclusive part of a preferably monogamous relationship (see Ernestine Friedl's book, *Sex the Invisible*). Others think our desire for privacy during sex comes from our moral brain. We are conscious of ourselves and of the consequences of our actions towards others. As with peeing and pooing, we know that sex can arouse hilarity and disgust within ourselves and observers. Bonobos will not make fools of each other when one of them has a crap, or two are having it off with each other. We do, which is why we do it with the toilet door and the curtains shut.

Farmer seeks incubator

The agricultural revolution caused great changes in the sphere of love twelve thousand years ago. One of its consequences was that we had more children, and men with high status in particular seized the opportunity to shackle several women to themselves and sire many offspring. It so happened that more stuffs, goods and warehouses with food appeared. There was property, something that had been inconceivable before. People began to collect goods and, in time, the amount of calories available to each person increased. These are all exaggerated cues that incite our brains to have many children. Whereas our ancestors two million years previously had on average two children reaching adult age per couple, the advent of agriculture led to more room for more tots. A man was able to have as many as ten children with one woman. Prosperity begot much larger families. The effect was an enormous population explosion. Because people no longer moved around with animals and the seasons, in places where there was farming, villages and later cities appeared. The property people had amassed had to be protected and fortunately there were plenty of people available to do this.

Power differences between individuals and families were small in ancestral times, because there were no possessions to fight over. Power relationships during the agricultural era changed fundamentally, however. The introduction of agriculture brought with it power relations humanity had not known for millions of years previously. Major landowners, kings, regents who transmuted their power into more children and extramarital affairs, but also wretched, suppressed and destitute individuals with little chance of offspring arrived on the scene. The familiar groups and compact kinships in which our ancestors had journeyed the land disappeared and were exchanged for top-down organisations, with all attendant consequences for love.

The effect on the different roles of men and women was dramatic. The more frequent pregnancies and larger throngs of children made women more dependent on men, and for much longer. Men produced food and goods, and provided protection against intruders. This reintroduced the issue of paternal uncertainty with a bang.

In ancestral times men and women were forever rubbing shoulders, which was just as well as the man did not have to doubt one of the million dollar questions of evolution: am I definitely the father of my children? This certainty existed, not for a full 100 per cent, but sufficiently. With the advent of agriculture and the attendant population explosion, the problem of paternal uncertainty became more topical, because in the agricultural environment the man would often work at some distance from where his wife and children were. The relative sexual freedoms of primeval times began to become problematic during the agricultural era. As a man, how could you be sure you were the father of the many children your wife had borne? One of the solutions to this pressing issue was the restriction of women's liberties. The egalitarian system that had functioned for millions of years was exchanged for the suppression of women and forced chastity. Sometimes, women were literally locked up at home or they were only allowed onto the street wearing a veil and chastity belt. A spouse was a personal breeder, no more than that – a mismatch with great consequences to female well-being. The question is why women allowed themselves to be restricted in their sexual choice and personal freedom. A possible explanation is that they accepted this in the interest of their offspring. Rather chained than repudiated and destitute.

A gene pool thrives in variation. Inbreeding enhances genetic defects or makes them more frequent. The advent of agriculture led to some cultures (for example Indian nations and areas in Bangladesh and Afghanistan) giving such importance to legacy that, against better judgment, people would rather marry a cousin than a relative stranger who might make off with the entire loot. Heirlooms and political rights had to be kept within the family at all cost. Certain genetic abnormalities continued to exist or even became more prevalent, with adverse consequences for the quality of offspring.

Whereas during ancestral times power and social differences between individuals or families were small, agriculture led to the emergence of men with greater means or status (major landowners, regents, kings) who seized the opportunity to secure more progeny by attaching several women to them by extramarital means or otherwise. On the other hand, wretched, suppressed and destitute individuals were very unlikely to produce offspring.

A greater intake of calories had another effect: girls started to menstruate at an ever younger age. For ancestral girls this was around the age of seventeen, while in 1955 the (estimated) average age was thirteen years and seven months. These days, girls have their first period roughly when they are twelve. The reason for this advance could be related to our changed eating patterns, wherein we eat a great deal more protein. In addition, we are said to ingest many more chemicals such as pesticides and plastics. The number of days an average child is sick has also plummeted. Research shows that smoking plays a part in the onset of menstruation; mothers who continued to light up during pregnancy have daughters who are more likely to menstruate early. A British study revealed that the average age for girls from socially disadvantaged families is twelve years and one month. For girls in more affluent families it is twelve years and six months.

This menstruation shift is likely to be a mismatch. In contrast to their yearning bodies, the young girls' brains are not yet ready for procreation. This may lead to poor partner choice. Differentiating between good and bad potential fathers for your children is a lot harder when you are fourteen than when you are eighteen. In 1995 the boy band Take That performed in the Netherlands; the daily newspaper *De Volkskrant* published shocking photographs of frenzied girls who came to scream at their idols. One of them, barely twelve if that, held up a sign saying 'Robbie, I wanna suck your dick!' Well, her English was flawless, that much can be said. Recently, bubblegum band One Direction gave a concert. Again the newspaper printed a photograph, this time with the text 'Point your erection in my direction'. The girl who held up the sign did not look a day older than thirteen.

Parasites

For thousands of centuries, the animal kingdom has been waging a permanent battle against parasites, life forms that keep themselves alive at the expense of other organisms. The influence of parasites on partner choice has changed, resulting in a potential mismatch. The battle between parasites and their hosts has shaped the vast majority of life on Earth. In this battle, the immune system of an organism was of vital importance. Their defence had to adapt swiftly in order to be able to ward off the parasites' terror attacks. But the problem was that parasites procreated too quickly, with their hereditary material changing at such a murderous speed, that for their hosts, natural selection would be too slow a mechanism to win the fight against their 'guests'. And this is where sex came into the fray!

Through sex, two organisms were able to merge their immune systems and in so doing ensure that they adapted to parasitic dangers much more quickly than through natural selection. Organisms look for organisms whose immune system is compatible with their own. In the battle against parasites quick action was of the essence, and sex was the engine to stay one step ahead of parasites. In short: someone who differs genetically from you would be an excellent partner, because your joint immune systems will be able to outstrip parasites. Sex is the best of both worlds. If someone has the same immune system as his or her partner, their descendants' immune system will not be enhanced, and so this not advisable.

Our genes are geared towards acquiring an attractive partner to make love to. By linking having offspring (is important) and sex (is enjoyable) evolution has found a neat solution and we do not have to think about whether we want to have children. It happens automatically. Until culture started to get involved. Breeding programmes with pandas had been failing for a long time without anyone understanding why. A whole mating programme had been set up to prevent inbreeding, but despite the fact that the bears were not related, the animals did not want to get in on the act. Further investigation revealed that this could be to do with the previously mentioned MHC genes which play a part in the immune system and odour recognition amongst other things.

Another example of mismatch is the use of the pill, introduced in the 1960s. Studies show that pill users tend to prefer men with comparable MHC genes as potential partners, which is not very smart from an evolutionary point of view. In other words: the hormones' preferences had been switched. It is likely that women who take the pill choose their partners in a different way to women with a natural menstrual cycle. The mechanism which ensures that a potential partner is sought whose defence mechanism against parasites (another definition of love) is compatible, can be disrupted by taking hormones. By using the pill, women no longer find some men's odour unpleasant. When, during a long relationship, it is decided to come off the pill in the hope of conceiving, this can have consequences as, in conjunction, both partners' defence mechanisms are not ideally matched. The pill plays into parasites' hands.

What's more, researchers at the University of Liverpool have found that genetically similar partners tend to have difficulty conceiving. And even if they are successful in getting pregnant, other problems arise: a greatly increased chance of miscarriage and of children being born with a weaker immune system.

The use of contraception makes our body 'think' that it is pregnant and that's why the eggs are being rejected. Everything related to a fertile period is suppressed. Studies shows that when they are ovulating, women think of sex more frequently, they may choose to dress more suggestively and in tests reveal an increased preference for more masculine men – when they are asked if they would go for a quickie, that is (with regards to long-term relationships there is no shift in attraction). American research shows that during ovulation women cast more left-wing and liberal votes, but we also know that women are more xenophobic when they are ovulating.

Evolutionary psychologists believe we are dealing with adaptive primal motives: during ovulation women would be more liberal as it would give them more sexual freedom and options, more xenophobic because of the fear that men from other groups would make them pregnant, only to sally forth again, and fall for masculine men as these would have better genes.

Relationships, research tells us, have a reduced chance of success if lovers met each other when the woman was taking the pill and subsequently stopped. If she goes for a less masculine man, then this man will sink below the minimum accepted level of masculinity once she stops. The feminisation of men, set in motion over the last thirty to forty years, could be the result of use of the pill. Another hypothesis is that women have become a great deal more right-wing by dint of contraception. The pill gives the female body a signal that it is pregnant and suggests it would be to the woman's advantage to have a decent, reliable, steady partner who will invest in her offspring. This strategy fits in with a conservative worldview of suburban bliss. American studies show that women who take the pill vote marginally more conservatively.

The use of alcohol, too, has far-reaching consequences for our reproduction. During ancestral times, it was usually only men who drank alcohol together, at set times and not all that excessively. Alcohol came from overripe fruits and was therefore not generally available. Present-day hunter-gatherers like the !Kung men and women do not drink their firewater together. Consuming alcohol is a religious ritual whereby men and women retreat separately into their tents. This can be seen as a plausible model for primeval times: alcohol was drunk sporadically and in a single sex environment.

Cereals cultivated following the agricultural revolution opened the tap. Nowadays alcohol is ubiquitous, and men and women drink like fish in each other's company. One of alcohol's many effects is that it impairs our judgment. It also lowers our moral barriers. Alcohol, Shakespeare wrote, 'provokes the desire', to lament in the same breath: 'but it takes away the performance'. Research shows that in a state of inebriation it does not take long before you find someone 'quite attractive, really'. People under the influence have sex more often, including one-night stands with partners they would never have dragged home or into an alleyway when sober. And sex produces children. Thanks to alcohol, progeny are being born that would not have been had it not been for some serious drinking. The question is whether the participants are then prepared to invest in these children.

Love Potion No. 9

Around 1600, many affluent Londoners had black teeth, because of an explosive increase in sugar consumption. People who were too poor to be able to afford sugar painted their teeth black to disguise their humble origins. Black teeth represented wealth and therefore 'attractive sexual partner'.

This great story by Bill Bryson about the time of Shakespeare shows how easy it is for us to be hoodwinked by the appearance of things. If a few centuries ago a dab of charcoal did the trick, these days billions are spent in the physical improvement industry. We mask our own scent behind litres of perfume, aftershave and deodorant. Everyone ends up smelling a lot more pleasantly, but the chance to determine someone's MHC compatibility has evaporated. Likewise, we inject entire lakes of Botox into us, have fat sucked out and are being remodelled with a vengeance. It's asking for mismatch.

Push-up bras, hair dye, eyelid correction, the sock-in-underpants, these have all been invented to fool the opposite sex, just like toupées, inflated lips, chin enlargements and all the other interventions to appear in better physical shape and therefore more attractive. As you already know, we can be quite taken by a healthy bosom, but if you allow yourself to be excited by blown-up, fake breasts or by a chest broadened by steroids, you will be deceived, not to speak of the resulting children. Plastic surgery is not genetically passed on, but a self-reinforcing meme: when two partners choose each other for their 'corrections', the child is more likely to have both 'imperfections'. In China, a man recently took his wife to court because she had not told him about the cosmetic surgery she had undergone before she knew him. He did not find out until he saw his children! (And although this is probably a hoax, it is still a good story.)

Women fall for men with status and money. Clothes are a status symbol that can enhance the man's attractiveness. As part of a study, women were shown a man in an Armani suit, a McDonald's uniform and a postal uniform. Women showed a preference for the man in the suit even though it was, unbeknownst to the women, the same man throughout. A suit gives off a signal that the man has a high status and it is to the woman's evolutionary advantage to choose this. Clothes are a means to increase attractiveness. Another study showed that women found men more attractive when they wore shirts with a Lacoste logo, even though all men wore exactly the same shirt – apart from the logo. Women fell for the logos.

Evolutionary psychology differentiates between costly and cheap signals. A Ferrari is manifestly a costly, honest signal: not every idiot can afford a car costing several tons. But in the case of clothes and watches, the brain can easily be taken in by fake versions. Just like body reconstructions. Pure mismatch.

Using gimmicks and behaving in a particular way, cougars (older women who have sexual relations with younger men) show that they want to appear younger than they actually are, whilst at the same time exuding maturity. This is a strategy to persuade men to choose them. When a man has the option to choose between two women, one of whom is definitely fertile and the other possibly not, 'maturity' can be a determining factor. Amongst chimpanzees, females who have produced offspring are one up on females who have not. A cougar, a hunting woman, gives off the signal that she is loose, an evolutionary red rag to men. But a (young) man who focuses on a beautiful older woman who is no longer fertile (but suggests she is), forfeits his chances to make an evolutionary contribution.

Nowadays there are men who take testosterone, especially body builders and older men. Their stomachs are said to shrink and their libido improve, as would the chance of their being able to beget offspring at a later age. Men are fertile until late in life, but at a more advanced age they are not necessarily better at investing in children. A recent Finnish study reveals that children of older fathers are more likely to have (mental) health problems (especially psychopathy). This could be the result of changes in the DNA that occur as men get older. Sadly for testosterone devotees and women who fall for them, sperm is not like old wine in new bottles.

Power and income are ways to acquire an attractive status, but pursuing it can end in mismatch. In modern society 'extended learning' is a vital way to obtain money and power. Extended studies can lead to the postponement of choosing a partner and coupled to this, the postponement of having children. This can result in highly-educated people being unable to convert their acquired status, income and power into reproductive success. A clear example of mismatch, because by postponing having children and rating this of secondary importance to education and career, their genetic heritage could die out. An issue in rich Western and Asiatic societies (Hong Kong and Singapore) is that people no longer opt to have many children, as was customary until recently, because of an increased focus on the quality of upbringing. Asian 'tiger mums' are an example of this.

One of us is involved in a big research project into 'child procrastination' in Singapore – one of the richest countries in the world with one of the lowest birth rates, 1.1 per two adults. In many affluent countries (including European ones such as Spain and Italy) the trend is increasingly for fewer children being born. In prehistory, a couple would raise two children to adulthood, now it is fewer than two because we (want to) have it so good. In affluent countries people are often involved in a status competition which means that both parties have to work to keep up the high-end lifestyle. The cue to 'get richer and become more successful than the neighbours' can lead to people postponing having children. It remains to be seen whether all that super food, those expensive schools, luxury holidays and infant yoga classes will lead to more offspring in subsequent generations – and this is the only Darwinist currency that ultimately counts.

What's more, these tend to be rich, densely populated countries. In nature, high population density frequently leads to increased competition for food, until a balance has once again been restored. This gives our brain and body the signal 'moderation in procreation!' The aim of the child procrastination project is to find out how the experience of Singapore as a busy, crowded environment can be altered so that couples may decide to have children sooner.

It's raining men

Thanks to the present (over)population, which is coupled to new ways of living together, many of the familiar, horizontal groups and clear family ties of our nomadic ancestors have disappeared. This has resulted in other kinds of mismatch. The more unstable or unsafe an environment in which a girl grows up, the sooner she will have her first baby. Biologists tend to see this effect occur in animal species when there is strong 'predator pressure': if a predator is about to strike, it would be better for the prey to pass on its genes swiftly than wait until it is too late. Studies in Chicago showed a direct link between the danger of a neighbourhood and the number of teenage mothers.

Research by the American development psychologist Bruce Ellis has revealed that – as mentioned before – girls who grow up without their biological father start their fertile period sooner. This could have several causes. Firstly, it could be purely genetic, of course. When a girl has a father who left her mother quite quickly, she might also be someone who aspires to short-term relationships. Many men go for the long term, the monogamous strategy, but others prefer the short term. A father who opts for short relationships could transfer this preference to his offspring. Psychologically, a girl without a father might be influenced by her mother who, as a single woman, also operates in the love market and in so doing subconsciously starts to compete with her daughter. Popular TV-programmes like *Hotter than my Daughter* are based on this principle. This is about competition within a household under the maxim: act quickly, otherwise my mother will be off with my potential partners. In terms of the four irresistible cues, competition with the mother could be an example of an obsolete cue to which the girl's brain nevertheless responds.

The almost unlimited number of people we are in contact with these days affects the current dating market, providing an enormous variation in genetic supply. This is essentially a match, but there are also disadvantages. Whereas our ancestors knew everyone's origins and kinship ties at tribal happenings, thereafter it was increasingly a grope in the dark. In a club – let alone on Tinder and on dating sites – the majority of people do not know each other, there are even visitors who have been born on a different continent. Knowing who you are dealing with is much trickier when you are setting up a chat with Martha from Mexico or Sean from Singapore. Is the more or less voluntary and impulsive choice of consorts without involvement of the extended family's knowledge and experience a reason for the many divorces in our culture? Amongst hunter-gatherers the figure for this is around 10 per cent, for us it is more than three times as high. Mixed marriages (between people of different cultures) are estimated to have an even greater chance of failure: one in two.

Another mechanism that may lead to mismatch is that the more men are interested in the same woman, the more attractive she is considered to be. We call this 'mate copying' and it also occurs amongst various bird species. Mate copying is the idea that an organism is guided by information from the environment in its choice of partner. When someone finds someone else attractive, third parties see that person in a different light. This is an ancient mechanism. We can try and check out everyone for ourselves, but finding out what others know about a particular person tends to be smarter. Yet this can lead to mismatch as well as we focus greatly on people who are unattainable, potential partners such as Justin Bieber or Rihanna. One of our daughters was literally sick with infatuation for Justin Bieber. When this singing guinea pig paid a visit to our country a 'meet and greet' was offered to interested teenagers. At a cost of four hundred euros per person eight girls would be allowed to spend ten minutes in Justin's presence for some photos and a chat. A ridiculously large investment into someone who is unattainable. It is not yet clear whether falling in love with inaccessible celebrities has a function. Until such a time we may safely consider it a mismatch.

Amor 3.0

Can we – cautiously – put on rose-tinted glasses to speculate about the future of love and sexuality? What is the future of these typically human traits in the light of what we know about human evolution and how our brain functions? How can we make the love mismatch a match again?

Love is . . . letting go. We should simply learn that we are living in times of love mismatch and accept that some people are more susceptible to this than others. The first group will continue to remodel and perfume themselves, take the pill and wait for the ideal partner until well over the age of thirty-five. The other group will cut back on alcohol, opt for non-hormonal contraception and start having children in good time.

An alternative scenario is futuristic but not inconceivable. Twenty years ago physical appearance was the only method to assess someone's genetic calibre; these days it is possible to have your entire DNA mapped for a small sum, and in future all kinds of techniques may exist that will be able to tell you how well your partner scores in genetic compatibility. In a few years' time we may be able to swipe left or right DNA profiles on Tinder, and perhaps everyone going out of an evening will do so clutching a chromosome passport. Perhaps we can reduce the harmful side-effects of divorce, pornography, alcohol and one-night stands by focusing more on how our etiquette functioned for centuries. Not out of fuming moralism or regressive conservatism, but in order to soften the excesses of our rapidly advancing cultural environment on our Stone Age brains. Shouldn't it be more normal in the West for parents to have a say in their children's choice of partner – and a little less in the East? In the end, *omnia vincit amor*, love conquers all. This is imprinted on our ancient brain after all. As Virgil wrote about this crazy little thing called love: *et nos cedamus amori*. Let us too yield to love.

CHAPTER 4

This isn't working

In cinematography, a tried and trusted technique is that of cross-cutting. *Four Weddings and Funeral* opens with a cross-cut of the morning rituals of the various members of a group of friends on the wedding day of one of them. This chapter begins with a cross-cut of three employees of the same company, who all travel to a vast office complex one morning.

The first shot features Thomas Akehurst, Head of Internal Logistic Planning. His external logistic planning leaves something to be desired this morning in the sense that, as usual, he is gridlocked in the daily traffic jam on the M4 motorway towards Thames Valley Business Park in Reading. He is firing off some emails to his colleagues on his smartphone from behind his steering wheel as yet another motorist leaves it to the very last minute to merge into his lane when the road narrows from three lanes to two, only just missing Thomas's estate car in the process. Thomas is furious.

On the A327, Nick Montague is cruising smoothly from his leafy villa in Finchampstead to the Reading head office in his Audi A8. His mobility manager – who a few years was simply his driver – glides the car along the road, allowing Nicholas to read the *Financial Times* on the backseat. Later that morning he has a conference call with some members of the Supervisory Board about the upcoming board meeting, but first he has a meeting with Thomas Akehurst, one of his longest-serving members of staff.

Meanwhile, in Reading, Josey Turner has just boarded the bus towards Thames Valley Park. She, too, is an old hand, having worked at the firm for exactly thirty-three years. When the new head office sprang up twenty years ago, her then boss asked her if she would like to exploit the toilet facilities along the public passage in the central atrium. Since then Josey has looked after the ladies' and gents' facilities, as well as five urinals, as if her life depended on it. She is sitting on the bus with mixed feelings, because today is her last working day.

Thomas Akehurst continues to wind himself up over other drivers in the traffic jam, whilst making exasperated phone calls with colleagues and the man he calls his 'personal underling'. There are some things that still need to be sorted out extremely urgently. Together with Maintenance & Housekeeping, an important logistic operation is being scheduled; it is so important that he could be judged on it for his year-end bonus. But this morning Thomas is first seeing Montague, an appointment that was made by the HR department. It kept Thomas awake all night, as he tossed and turned over the question of what his CEO and grand inquisitors wanted from him.

When Nick Montague arrives in the central atrium, he remembers that his executive assistant Maryolyn – whom he simply called his secretary a few years ago – had urged him to call in on the lavatory attendant. Shortly before he left yesterday, Maryolyn had even given him an envelope for the woman, as a modest thank you from him for all those wonderful years of public peeing pleasure.

Josey has meanwhile put her patch in order, all the sanitary ware is clean and all soap, towels and toilet rolls have been replenished. Her saucer for payment is on the table, next to a large bunch of flowers she has been given by the staff association. She may not be a member of staff officially, but everyone here looks on her as a family member. The agreement she made with the then Household Services Department was that she would end her employment in exchange for full management of the lavatories. The more people peed, the more pennies Josey would earn.

At nine thirty, with a heavy heart, Thomas Akehurst reports to Nick Montague's executive assistant. He has to spell his name twice before she is able to locate him in the appointment diary. Thomas has to wait for eighteen minutes on a couch in the passage until he is allowed to enter his boss's office. He has hardly set foot in it during the past few years. Montague and Thomas's line manager await him.

Exactly twenty-three minutes later Thomas leaves the boardroom, escorted by a staff member from Corporate Security. They have assured him this is not personal, but purely company policy. Montague was amicable, yet correct. Akehurst's department is going to be abolished and amalgamated into a new joint Logistics Management Operations Unit, which means that, regrettably, after all these years of loyal service Thomas has become redundant. His duties will be taken over by his underling. Thomas must leave the premises under supervision, as usually happens in American films. Montague thanked him for the years he has devoted to the firm and assured him that the company will make every effort to ensure he will find a new job before long.

Clutching a cardboard box with his personal belongings Thomas makes his way through the central lobby, accompanied by one of the security guys. Josey spots him walking past and gives him a cheery wave. Thomas puts his box down and enters her territory for one last time. Having peed, he points to the fresh flowers next to the saucer.

'Got something to celebrate?'

'I'm retiring today!' Josey exclaims. Thomas gives her a hearty, yet glum look.

'I hope the company took good care of you.'

Josey laughs and tells him that they have not done anything of the kind. She is her own boss. And then she drops her voice. 'I've never told anyone here before, but now that I'm leaving I can say it. At first I was afraid I wouldn't earn nearly enough on my pin money income as a freelancer, but after the first year it turned out I didn't do too badly. All that small change brings in about three hundred thousand pounds a year gross. Some days my daughter and I would go home with a thousand pounds! The one thing I don't need is a pension!'

'That's good,' Thomas replies, but in a tone that suggests he is not really listening. He puts a fifty pence coin on the saucer and wishes Josey all the best.

On the M4 back to Newbury there is no tailback.

The original freelancer

In a sketch from the television series *The Armstrong and Miller Show* a caveman is sitting opposite three members of another tribe in a job interview.

'What do you do?' asks one of the tribe members.

The job applicant replies: 'Me hunter.'

'We have many hunters,' the female tribe member exclaims.

'Me also gather!'

Third tribe member: 'You hunter-gatherer?'

The job applicant replies proudly: 'These days man need many skills!'

In ancestral times, work did not exist as a concept. Just as chimpanzees do not have jobs or worry about their salary slips, prehistoric humans had no notion of employment. Everyone in the tribe had to forage for food and help prepare it, everyone was responsible for keeping the fire burning, everyone helped defend against enemies and everyone looked after the children, both their own and those of other tribe members. Once our ancestors had satisfied their basic needs they would engage in social activities, such as storytelling, discussing politics, singing, dancing and other group-bonding rituals. There was no distinction between 'work' and 'private life'. There were no bosses, jobs, contracts, salaries or pensions. In short, ancestral humans were all-round freelancers.

Scientists have long thought that, most of the time, hunter-gatherers were caught up in a harsh struggle for survival and that starvation was never far off. In the 1960s this picture was radically revised by anthropologists Marshall Sahlins and Richard B. Lee. They came up with the theory of the original affluent society, which assumed that hunter-gatherers led a life in which their needs were comfortably met and that they did not suffer at all. On average, ancestral people needed to put in far fewer 'working hours' to stay alive than people nowadays, and much less still than compared to the agricultural and industrial revolutions. According to these anthropologists, our ancestors had a 'marvellously varied diet' and lived in a world of 'affluence without abundance'. Once they had enough, they had enough. An earthly paradise, in other words.

Studies of contemporary hunter-gatherer peoples like the !Kung and the Hadza show that they only spend fifteen to twenty hours a week on what we would call work. The rest of the time they engage in what is described as 'social interaction'. They laze around together, get in touch with their ancestors using ancient rites, discuss and celebrate life. 'Later' does not enter the picture, because later is a meaningless concept.

Early *Homo sapiens* was literally *Homo universalis* (or all-round freelancer) and this is still the case amongst today's hunter-gatherers. There were of course some minor specialisations: broadly speaking, it was the men that hunted and defended the group and the women who collected nuts and fruits and looked after the children, but these tasks indubitably overlapped. People mainly did what they were good at, especially if it benefited themselves and the group. But often they also executed tasks they were not skilled at, simply because these things needed to be done in order to stay alive. People needed to eat, tidy up, move on and find a place to sleep. Not unlike a modern camping trip, where the family members all have to pitch in, without being able to rely on supermarkets, landlords or plumbers.

In prehistory, even the most influential and prestigious person in the tribe, the leader of the group, spent the entire day doing what was necessary to look after himself and his direct environment. In his spare time he would be involved in group politics, a bit like the chairman of an amateur football club or the head of a motorbike club today.

Our ancestors 'worked' close to the place where they would sleep that night. They did not have to commute or face traffic jams. When the men of the tribe went out on a hunt they might be away for a few days, especially if prey was some distance away, but they did not travel far on a daily basis. Friends, relatives, colleagues and fellow tribe members: all these categories merged, and everyone was part of a greater whole. There were no actual colleagues, no clear distinction – either physically or psychologically – between private life and work. No one needed a CV, no one needed to study or train or plan their career. Youngsters learnt skills 'on the job' from adults, not unlike in the successful television series *The Apprentice*, in which Donald Trump (or in Britain, Alan Sugar) offered talented young people the chance to become his apprentice. If you wanted to become a hunter, you would just join in with the best hunters and learn the tricks of the trade from them. Developing talents in certain valued fields gave you more prestige, and ultimately more sex and offspring.

Bartering leads to wealth

In ancestral times, some bartering took place and this produced tentative specialisations. If you came back from the savannah with a large chunk of meat, you could swap this for other things, sex, for instance (the previously mentioned 'meat for sex theory'), or an extra supply of firewood. Trading is in our genes. The British economist Adam Smith (1723–1790) argued in *The Wealth of Nations* that people have an urge to exchange things, a tendency that has not been observed in any other animal species. In his famous words: 'Man is an animal that makes bargains: no other animal does this – no dog exchanges bones with another.' We do, and this started in the depths of prehistory. In Australia, researchers have found exchange networks stretching thousands of kilometres, whereby shells were found in southern Australia that came from beaches in the northern part of the country. Conversely, spearheads were discovered in the north which had been made using stone from the south. So a trade route clearly existed in which valuable objects were exchanged for others.

This is called a 'barter economy'. Before the invention of money, bartering was the only method of trading. But there are many snags attached to this type of economy. Firstly, the demand for things and the supply of others is not always in sync (imagine you have just eaten and someone calls around with a piece of meat to barter for a spearhead you have been working on for the past few days); secondly, many things are perishable (meat starts to smell rather strange pretty quickly in the absence of a fridge); thirdly, it is difficult to assess the exact value of goods (why would one spearhead be worth two chicken legs and not three?); and finally, not all things are easily dividable (it is difficult to swap half a spearhead with someone).

Looking for cheaters

Evolutionary psychologists Leda Cosmides and John Tooby argue that the human brain has adapted to trading and bartering as it offers advantages to the individual. 'I want something from you and I will reward you for this now or later.' We call this the reciprocity principle. In order for this system to function effectively, people need to have the ability to recognise and expose potential cheaters. This 'cheater detection' mechanism exists to prevent us being exploited during a barter. The mechanism looks for information that can predict or reveal whether the person we are trading with is taking us for a ride. Work it out for yourself: who are the people you would rather not give your wallet to? Who are the people you would rather not give your house keys to? You will probably find there are only a small number of people you trust implicitly.

In order to ascertain whether people were able to detect a cheater, Cosmides gave students at Harvard University the so-called 'Wason selection task'. This is a logical puzzle introduced in 1966 by cognitive psychologist Peter Cathcart Wason. People are shown four cards with which they have to prove a proposition. Say you see the following four cards on the table in front of you:

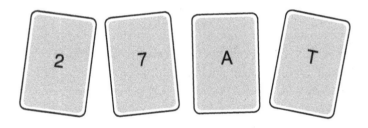

The proposition that has to be proved is as follows: 'If the card has an even number on one side then the opposite side will have a vowel.' Which card do you need to turn over to prove this proposition is correct? The answer is ... (Try it out for yourself.)

The card with the even number (so 2) and the consonant (so T). The first action should show a vowel and the second an odd number for the proposition to be correct.

Don't worry: fewer than 10 per cent of the subjects got it right. It turns out we have great difficulty with this kind of abstract logic. That is ... until we replace the abstract standard concepts with examples from our daily life and our normal conduct, especially when they relate to detecting cheaters. Then it is a lot easier for us to understand the logic. When we take these cards for instance:

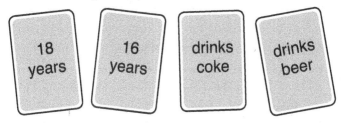

and the proposition states: 'If you are drinking alcohol, you must be over eighteen.' The question is again: which card should we check? Card number one says 18 years, so does not have to be checked. Card number two says 16 years, so is not allowed to drink beer. If we turn over this card and if the reverse shows a sixteen-year-old with a bottle of Grolsch in his hand, than we caught him out. Card number three drinks Coca Cola, so does not have to be turned over, because anyone is allowed to drink coke. Card number four drinks beer, so we need to check this one to see if the beer drinker is over eighteen.

The majority of people had no difficulty with the solution turn over cards '16' and 'beer'. According to Cosmides and Tooby, applying these rules of logic is one of the mechanisms with which we can expose cheaters. When someone lends out their car under the condition that it has a full tank of petrol on return, it can easily be checked whether someone has stuck to thus condition. If the tank is not full, then the car will never be lent out again.

It appears that, in a social context, people are able to reason much more rationally than in other areas and this makes it likely it is an adaptation for reducing the chance of being cheated – especially in the context of exchange and favouring. We were quite happy to share extra meat with a fellow tribesperson, but only when we got something of equal value in return.

How does the Stone Age affect our trading brain?

We still like to swap, from stamps to football cards, and by doing this we give value to things as we did in ancestral times when money did not yet exist. Just as we did, our sons collect Panini football stickers which have become a veritable cult. Newspaper reports suggest the Italian family firm has distributed more than 100 billion cards since 1961 with an additional five billion in sixty countries being added annually. When, in 2014, Panini once again released special collections for the World Cup in Brazil, a reporter from *The Economist* calculated that ten people would jointly have to buy just over fourteen hundred packs of five cards to complete all ten collection albums. No child or parent can afford to be buy fourteen hundred packs, so if you want to complete your album you will have to swap 'like a caveman'.

We also continue to get a great deal of satisfaction in lending each other things: 'Hey neighbour, can I borrow your hammer drill for a bit?' There seems to be a new need for this barter economy in society. Many new community initiatives have sprung up, partly assisted by the internet, through which individuals can borrow things, swap or share objects or services with each other. On a site like choreswap. com someone can offer a particular service ('I am a childminder with ten years' experience') in exchange for another service ('I need a high-pressure cleaner for this afternoon'). Sites like Peerby and Konnektid allow neighbours to borrow things from each other. It does not involve any money and the tax authorities do not get a look-in. Nothing primeval is alien to us.

Talking fosters exchange, trade, negotiation and acts to prevent abuse. From an evolutionary point of view, talking about others is central to human social relations. In earlier times, talking about other people ('gossiping') was a way to increase group bonding and to test someone's reputation ('Can I trust him or is he a cheater?'). What happens around the water cooler, or outside on the pavement during a cigarette break, took place around the campfire then. Evolutionary psychologist Bram Buunk wrote an instructive and amusing book called *Primitive Drives in the Workplace* (*Oerdriften op de werkvloer,* 2010) about the significance of gossip at work, amongst other things. He shows that in most of the scientific literature on organisations, gossiping is viewed as something negative, while it is not, or at least does not have to be.

Some scientists, like evolutionary psychologist Robin Dunbar, believe that language evolved in humans in order to be able to survive in large social groups. Humans need fellow humans to survive. Whereas apes and humanoids remain on good terms with each other by grooming, stroking and tickling, humans chat to start a social relationship. Language allows you to attend to more people at once than grooming, so using language, you can live in far bigger groups. We gather information about the group through language and we can share our opinions, desires and irritations with others. We are the Talking Ape. But we are also the Gossiping Ape.

Buunk shows that gossiping has several functions. Our brain is not really equipped to converse in very large groups, and as soon as a group becomes too big, the conversation splits into different smaller exchanges. Check this out when you are next at a party or in a meeting. When more than four people are engaged in a conversation, it breaks up into groups of two and three unless there is a chairperson. When exchanging gossip, the objective is not so much to share information as to share opinions. Extensive gossiping is 'a kind of investigation into the others' deeper motives, where their allegiance lies, and what their intentions are'. And because we tend to remember negative tales better than positive ones, 'gossip' has a nasty connotation. Yet malicious gossip is much less prevalent than neutral or positive talking-behind-peoples'-backs. Research by Robin Dunbar and others shows that as little as 3 to 4 per cent of gossip amounts to true defamation.

In her book *Gossiping, why Gossiping is Healthy* (*Roddelen, waarom roddelen gezond is*, 2006) Flemish communication scientist Charlotte De Backer distinguishes no less than sixteen categories of gossip. The function of gossip is largely to gather information about relevant evolutionary goals we pursue: about what we can learn (1) about our health, (2) finding and keeping a partner, (3) contact with other people, (4) the reputation of others, (5) who is having sexual relations with whom, (6) who our sexual rivals are, (7) whom we can demean, (8) how we can detect cheaters, (9) with whom we can cooperate, (10) or not cooperate, (11) to whom we are related, (12) who could be an ally, (13) who might be our enemies, (14) how we can keep up or enhance our reputation, and finally, (15) how we can refine our image of others vis-à-vis third parties and (16) ourselves.

Besides information-gathering, De Backer distinguishes three further important functions of gossip. Gossiping is also an instrument to keep tabs on others ('Did you know that John sent an email to the manager about you?'), to manipulate others ('If I were you I wouldn't trust John, because you don't know what he wrote in his last email') and – last but not least – it is a form of entertainment.

Traces of our ancestral brain can also be found in the selection procedures for jobs and training we use these days. Nowadays, we have standardised tests to measure someone's intelligence and competencies, as well as assessments to determine whether employees fit in with a company or not, tests that obviously did not exist in the past. But even if such a test could predict with 100 per cent accuracy that someone would be a good match for a particular organisation, would you take him or her on based on that alone? No, of course not. We want to see the person before we take them on, we want to speak with this person, hear or even smell them. An (interesting) clue is that selection agencies who screen job applicants for other companies, often do not conduct tests when recruiting their own staff. Even occupational and organisational psychology departments – often the creators of these tests and assessments – are guilty of this.

Other ancestral traces: Israeli researchers have recently found out why we shake hands with (new) colleagues. When scientists secretly filmed people who shook each other's hand, many were shown to be sniffing their own hand subconsciously afterwards (there is some great footage on the internet of this). The precise mechanism has not been fully identified, but the hypothesis is that hand odours can impart information to our brain about what emotions someone is feeling: if the smell is 'anxious' then this might indicate that the other person is hiding something. Bear this in mind when you next shake someone's hand.

As we spend so much time at work in modern times, our Stone Age brain wrong-foots us by seeing our colleagues as intimate friends ('intimates'). Nowadays, we have a life at home and a life at work. In the past this separation did not exist, and the fact that we spend so much time at work now causes many problems in our intimate relationships. If you saw someone a great deal then it would have been a sign of intimacy in times past. Seeing a great deal of someone in effect meant that someone was close to you. It was relatively easily to avoid someone we did not like by moving to a different camp, for instance. Intimacy was a free choice. When, day in day out, we are forced to spend time with people in one single space our brain interprets this as a sign of intimacy. And so many workers confuse their working environment with their home environment. Our brain is primitive; if we see someone every day, it is a sign that we simply have to care about him or her. And thus we fall in love with our secretaries and we get cross with our colleagues for fixing up a meeting with someone else. Studies show that over 30 per cent of workers have been in love with a colleague at some point. Around 18 per cent of workers are said to have been in a relationship with a colleague.

Farmer seeks gate-lock lubricator

The agricultural revolution and the much later industrialisation led to many work specialisations. Humans turned from self-employed workers into specialised wage slaves. Many people lived in close proximity to each other and food was readily available. What had started out as bartering in prehistory became an efficient system in which one person produced cereal (the farmer), another turned this into bread (the baker), and yet another exchanged his meat (the butcher) for this bread. And so everyone's interest was served. In his renowned book *The Wealth of Nations* (1776) Adam Smith wrote: 'It is not from the benevolence of the butcher, brewer, or the baker that we expect our dinner, but from their regard to their own interest.'

The economist David Ricardo (1772–1823) was one of the first people to study the economic interest of specialisation, not only on countries, but also on individual workers ('the theory of relative comparative advantage'). By specialising, people, regions and countries were able to organise their labour more efficiently, which in turn produced more opportunities for exchange. Again, we can see this in the modern microbrewer Oedipus in Amsterdam, set up by four friends in the slipstream of a wave of new microbreweries. Initially all four did everything together and they spent more time with each other than with their mothers. The four young men – a psychologist, a hydrologist, an anthropologist and an artist – knew each other from school. They brewed their first beer literally in an attic kitchen. Their beers with lemongrass and Sichuan peppers were such a hit with the public that soon they were brewing 800 hectolitres a year. By then the niches in their company had been filled. One friend simply turned out to be the best brewer, the other the best salesman, the third was most at home in financial administration and the fourth came up with the best plans for the future. Specialisation was the catalyst for their success.

This is exactly what happened after the invention of agriculture; people began to fill social niches. Soldiers, civil servants, managers, seamstresses, weapon manufacturers, builders, fishermen, bakers, metalworkers, etc. appeared and ultimately all possible positions were taken by people with particular talents. The more complex a society was, the better it was for specialisation. The only thing the system lacked was something to allow the farmer to trade his goods relatively easily with the products of the butcher. And so transaction and trade agreements had to be registered in a memory you could hold in your hands. The first script to be found (dating back around five to six thousand years) is a note about a business transaction. The agricultural revolution ultimately saw to it that we developed ways to store and convey information. The fact that you can read this now is one of the many consequences of our liking for bread.

The increase in work and specialisation also sparked a need for organised education. In past times people used to follow one of several role models in the tribe on a daily basis, but this no longer sufficed in more specialised times. Before anyone was able to beat metal or engrave inscriptions they would have had years of instruction. Schools were set up, guilds created, rules and school legislation introduced. When all social niches had been taken up, it produced a relatively stable cohesion. Thanks to succession – another phenomenon that barely existed, if at all, in prehistory – the son of the butcher automatically became a butcher and not a jeweller.

With the growth in specialisations and the increasingly lengthening trade routes a need arose for the fishmonger to be able to trade his fresh produce for the services of the midwife relatively smoothly. A way to determine the value of specialisations had to be introduced, as 'bartering' and 'commodity money' often created problems around divisibility and shelf life. Money began to be used independently some 2,500 to 2,700 years ago in three different places (in China, India and around the Mediterranean). The introduction of money as a payment and barter method generated a great deal of new work in itself, because money had to be minted, administrated, guarded and transported. So money is a recent invention, which is why our brain finds it difficult to assess its value. What can we do with a million pounds, how much does a pint of milk cost, what did you pay for this book, how much do you earn each year after tax? Multiple studies show that people have trouble handling money. This makes total sense, because money is not a biological adaptation, but a recent cultural innovation. Money is a fake cue. A mismatch is lying in wait, in other words.

While the agricultural revolution produced the farmer, the industrial revolution produced 'the labourer'. The steam engine changed the production process in countless numbers of ways, leading to hyper-specialisation. Workers became entirely alienated from the products they made. Labourers worked on tiny cogs in the machine. From textile production to clothes to vehicles, the working classes were deployed to do mind-numbing repetitive work in large-scale factories. Things were done on an unprecedentedly huge scale, which only increased trade, which in turn created innumerable new problems.

Next to the factories, large, gloomy residential districts were built to house the workers. Society became divided between the haves and have-nots. Capitalism arrived on the scene, and to counter this, socialism, which tried to turn the world away from what was called 'alienation': people and the labour they produced grew apart in a way that was unprecedented in our evolutionary history. The industrial revolution has not finished yet, we are still slap-bang in the middle of it, even if we call it something else now: the digital revolution. We are living in a society in which galloping technological developments continue to have a large influence on work processes (for example computerisation, digitalisation and internetisation during the past decades). The result is a whole lot of new cues that mislead our brains, as well as mismatch.

Mismatch and work

Mismatch arises when our primitive brains put us on the wrong foot by responding to cues that – in ancestral times – would have led to behaviour that was beneficial for us. But in the work environment that we have created for ourselves, various cues cause us to make the wrong choices, with adverse consequences for our well-being.

To begin with, we find it hard to accept the fact that we work for a boss. Put simply, we cannot stand our bosses. As a result, we are not all that motivated to do our best. The majority of employees arrive at work promptly at nine in the morning, to leave on the dot of five. If we wake up with a hangover or a dormant headache we have no compunction in calling in sick. This behaviour stems from dissatisfaction with our 'work–life balance'. That's not what we were used to in ancestral times, because we were our own boss back then.

Organisational climate surveys in companies show that people are most stressed by interaction with their direct boss, their line manager. In the savannah, there were no managers or middle managers. Decisions were taken by the group, on the basis of consensus, not on the basis of hierarchy. Modern organisations have become excessively formalised and institutionalised, which goes against our small group instincts. Studies show that employees need a great deal of autonomy, a primeval preference for self-employment. People want to be left alone, they do not want some process supervisor breathing down their neck. The same studies reveal that employees consider autonomy and social contacts more important than pay. That, too, is an ancestral preference. Our desires have not changed, only the circumstances in which we operate.

Specialisation

You probably do not give it a moment's thought, but the letters you are reading have been designed by someone and this person has been paid by the publisher of this book (we are talking about a sum of 150 dollars, to be precise). This typeface is called 'Granjon' and was designed by George W. Jones in the period 1928–1929.

In the period immediately following the invention of printing, lead letters were cast by the printers themselves. After a while, specialised type-founders emerged. Typesetting was done by hand: compositors picked the individual letters from a type case and set these, upside down, in the correct order in a metal composing stick.

From the nineteenth century onwards manual typesetting was replaced by automatic typesetting machines. Specialised type-founders continued to supply letters to printers without their own typesetting machines right up until the 1970s. Specialised type-founders are now past history.

Mechanisation has also made the printing process less labour intensive. During the beginning of the twentieth century, the first offset presses were developed and printers now use computer controlled laser techniques and systems, leading to the demise of many jobs. It should be evident that there are many mismatches associated with too much specialisation.

Take one of the authors of this book. He is a tiny link in an enormous machine called 'science', the only expert in the world in one particular discipline. And yet he keeps all kinds of people around him in work, his house clean, his children educated, his food prepared and his finances in good order. Where can he go with his specialised education and profession if his university decides to make cuts in his department? Strong specialisations are a mismatch, because when even the tiniest thing in the environment changes the result can be pretty catastrophic. The entire house of cards can collapse (as the recent global economic crisis was likewise the result of a cog that got jammed: too little liquidity in the banks).

Specialisation has led to alienation at work. More than anything else we like to do work that is worthwhile. We want to be involved in 'the bigger story'. But in most organisations this cue is missing. Someone who is responsible for a tiny part in the production process loses overview and involvement. Our autonomous brain 'does not like' being a small cog in a much larger wheel. This has been described with compelling beauty and insight in the Dutch fictional cycle *The Institute* (*Het Bureau*) by J. J. Voskuil, an epic running to over five thousand pages about an institute engaged in the recording of information, with 'folklore' being the protagonist's specific responsibility. The book tells the story of a solitary individual working in an office, faced with fairly pointless work, office humour, buried tensions, irritations, harassments and the obsequious behaviour of colleagues. As the cycle recounts the story over a period of thirty years, readers see what being in a single enclosed office and trying to do your work day after day, year after year, with all the ensuing suppressed emotions, does to you.

Stress

The dividing line between work and leisure was an issue for people as far back as the mid-nineteenth century, whereby happiness was defined as having as little separation as possible between your work and your play. The concept of 'work–life balance' is comparatively new, dating back to the late 1970s in the UK and used only as recently as in 1986 in the US. Many studies show that this balance is often out of kilter, tipping too much towards work, to the great dissatisfaction of many employees. We want days off, which shows where our priorities lie.

As mentioned, our brain struggles at various levels with the way in which private life and work intermingle. A mismatch occurs because we confuse an intense working relationship with intimacy and love. Since we tend to see more of our colleagues than our husband or wife, this forms a constant threat to our relationship with our partner. That's why some organisations ban romantic relationships between colleagues, and in the event of two co-workers becoming an item, one of them has to leave. The workplace is often a hierarchical environment, which also makes intimacy risky. Some American universities dismiss teaching staff if they start a relationship with a student.

Our primeval brains cannot distinguish between work and private life. That's why a lot of us suffer from work-related insomnia – we take work home with us. Recent studies show that insomnia does not exist amongst hunter-gatherers. Work can take a huge toll on family or marriage. People experiencing tension in their relationship tend to ascribe this largely to their work situation. Seventy-five per cent of people come home late from work, 72 per cent feel tired because of work pressure and 48 per cent think they spend too little time at home. We have separated the home and work domains even though, from an evolutionary point of view, they are inextricably bound-up.

This is connected to how we experience stress in the workplace. In ancestral times, the threats we faced tended to be brief: a snake rustling in the undergrowth, a predator that had to be avoided, an attacking enemy. In these situations our stress systems would be activated, cortisol would be produced, and we would make a choice between fight, flight or freeze. This response was vital; without it early humans would not have survived. Stress lasted only a short while and when the threat had been warded off, our systems would recover their balance. The workplace causes a completely different kind of stress, as neurobiologist Robert M. Sapolsky of Stanford University explains in his book *Why Zebras Don't Get Ulcers* (1994). Work-related stress keeps our stress systems in a continuous state of readiness. The fact that our bodies are constantly awash with cortisol causes all kinds of health problems, problems that do not affect zebras (or ancestral humans). Much of the tension we experience is directly or indirectly related to work, and it is long term. Our bodies are not set up to process an incessant supply of cortisol, leading to burnouts, ulcers and heart disease.

In the most extreme cases stress makes the workplace literally lethal. According to statistics from the US Department of Labor, on average over five hundred employees a year were killed at work between 2006 and 2010, the majority as a result of shootings. Violent robberies were the main cause, but there were also instances where employees turned their gun on their colleagues. Managers are more at risk, as became evident in October 2015 when angry Air France staff physically attacked some of their bosses in protest at the loss of thousands of jobs.

Remuneration and retirement

Our present reward system, involving money for services, is a recent invention with in its wake a multitude of irresistible, exaggerated and occasionally perverse cues. Our brain has great difficulty assessing the value of money. Our ancestors may not have been involved in salary negotiations or pension planning, but the question has not changed: is the exchange fair and is it enough? But, these days, how do you respond to the question about how much you really need, whether it is fair and what you should do in return? We look at what colleagues earn and want to earn something similar, or rather a little bit more.

Ancestral humans ('freelancers without a pension') were unconcerned with provision for old age, and their modern counter-parts are likewise suffering from a persistent lack of interest in pensions. The mismatch here is that, with our long life expectancy, we would benefit from a decent pension but this is something we cannot grasp with our prehistoric brain. We have to release some money in the here and now for an uncertain future which instinctively feels a long way off. We can postpone building up a pension every day. Investing in a pension is rather like stopping smoking: you can keep putting it off . . . until it is too late.

Over the past few decades the labour market has changed fundamentally. Up until about twenty years ago it was organised in such a way that – apart from an assortment of entrepreneurs, independent traders and professionals – many were in full-time employment and worked long-term for one single employer. A pension was a sum that was deducted on a monthly basis from your salary. People did not think about their pensions. They assumed that the State would look after them, even when they had stopped working. Over a period of twenty years the labour market changed materially. There has been an increase in freelance workers, the welfare state was overhauled and working for a company no longer meant security. In the Netherlands, there were almost two million self-employed people on zero-hour contracts at the time of writing. According to the latest figures, in the UK 4.6 million people are self-employed, which amounts to 15 per cent of the total workforce.

The need for a decent pension has become much more pressing, but people's interest in it seems to have diminished in direct proportion. We see here mass cognitive dissonance and avoidance behaviour. In addition, there is major dissatisfaction with the financial sector, which is linked to the reprehensible conduct by banks and insurance companies. Many people have been the victim of misleading and over-priced insurance policies and shady financial constructions. This begs the anxious question: Will my money still be there when I need it in thirty years' time, or will it have lost all its value due to hidden costs?

The pension crisis is the result of mismatch. If self-employed and zero-hour contract workers are not planning on setting up a decent pension we are heading for a massive social problem. There are simply not enough bridges for all those future elderly to sleep under (or jump off). Traditional marketing has great difficulty in getting through to the thirty to fifty age group. The fact is that, for years, pension companies have painted a Fata Morgana of a mellow post-work lifestyle; a pension was something we needed when we wanted to travel the world as sprightly, cheery sexagenarians. Nobody has had the courage to produce negative ads about what will happen to us when we do not look after ourselves.

Lazybones

Workers suffer many physical complaints that are related to mismatches at work. We spend too much time sitting in the car and we do our work sitting down. This means we do not move enough, and this is bad for us. All kinds of studies have shown how beneficial exercise is. Physical activity is *the* way to increase creativity, productivity, learning and decisiveness. According to movement scientist Erik Scherder, Professor at the VU University of Amsterdam, the elderly in particular should exercise more, because this is good for their motor system and above all their memory. The latter has to do with the fact that new neurons are produced during physical activity.

Workers who engage in some vigorous physical exercise before work turn out to be as much as 23 per cent more productive, research shows (although cause and effect are difficult to separate, because maybe these are simply extremely fit people who enjoy working up a good sweat before they get going on their day's work). Scientists compared workers who go to work by bicycle one day and by car the next. Cycling for thirty minutes was enough to have a demonstrable effect on creativity and clear thinking, and this effect continued to make itself felt some hours after the exertion. In other words, organisations should encourage bicycle use.

Lastly, difficult decisions turn out to be a lot less difficult when workers move around more. Following a lengthy study of executives and staff it emerged that more physical activity led to a dramatic improvement in resolving difficult issues. A further study showed that fit, active office workers make 27 per cent fewer mistakes than their sedentary lazybones colleagues.

Together we'll find a solution

We can see various ways in which we can make a contribution towards dealing with mismatch problems in the work place. Obviously, we could begin by doing nothing, this is always an option: sit by and wait for things to run their course. People so obsessively preoccupied with their work have no time to have children anyhow and thus remove themselves from the gene pool, in favour of less ambitious colleagues. The problem will solve itself.

But to do something and to give our work more purpose and meaning, we could tackle the extreme specialisations. A working human being would like to have affirmation that his efforts are worthwhile. Assembly line work is found to be increasingly unpleasant in this day and age and people want 'social value' when carrying out their duties. People no longer want to be a cog in a much bigger machine, which is reflected in the production processes of companies like Volkswagen. Systems there have been changed so that every employee is responsible for every part in a car and can be employed to assemble any part. They no longer produce this one door, but the entire car. Today they are working on this particular door, tomorrow on a window, the day after tomorrow on front wheel valves, etc. Work must be worthwhile. And it works: since the introduction of these working methods Volkswagen employees have enjoyed going to work a lot more.

According to organisational scientists, organisations are increasingly being populated by motivated professionals, who should not be regulated and monitored too much. A collective ambition seems to be a much better mainspring than managers who 'try to plan work processes'. Workers flourish when they are actively involved and do not always have the same boss above them. They want to be able to have a say about how they work, and under which circumstances. Works councils, participation councils and internal consultation forums are matrixes that fall back on the ancestral organisation of the tribe, in which everyone's input was guaranteed. Co-determination is a match.

Along the same lines is the creation of smaller, flatter organisations and company divisions which reduce anonymity within companies. The size of our neocortex, the thinking part of our brain, determines how much social information we are able to process and how large these groups should be for us to feel comfortable (this is the social brain hypothesis we encountered in Chapter 1). We thrive in groups of one hundred to one hundred and fifty people. It is impossible for us to apprehend companies that are too big for our social brains.

In order for us to be able to function with our ancient social brain we could introduce a tribe culture within every layer of a large company's hierarchy. This means that monumental organisations such as Philips or the University of London should consist of units of approximately one hundred to one hundred and fifty individuals, in which staff members interact on an egalitarian and informal basis. Above these different departments would be an over-arching supervisory department, also made up of a 'tribe' of around a hundred people. Thus, an enormous organisation could be reconstructed from the bottom up.

An example of a huge company with a flat organisational structure is the American Gore & Associates, that produces a large variety of synthetic fabrics, such as the famous Gore-Tex. The company was set up in 1958 by the chemist Bill Gore and his wife Vieve. It now employs ten thousand people, but the amazing thing is that they do this without managers, without job titles and with an extremely simple structure. If one unit grows larger than one hundred and fifty workers, a new department is set up that does exactly the same work. Everything in the company is focused on personal relationships and social interaction. Gore is structured in small units that are responsible for a particular segment or process. Every team has a leader, but he or she is chosen by the group itself on the basis of questions and requirements which were also asked by our ancestors: 'Who should I follow?', 'Who is best able to help me?' and 'Who will teach me the most?' The Brazilian company Semco likewise works according to this ancestral philosophy and is also successful. Network governance is a method to cancel out mismatch. Some governance experts believe that this is the organisational model of the future, but in fact it stems from our distant past.

Because of kin selection, our ancestors had a great deal of confidence in their relatives, and this has repercussions in the present day. In the Netherlands, there are – depending on how it is defined – around 180,000 family firms who jointly are responsible for over 40 per cent of jobs and almost 50 per cent of GDP. Of the big companies with more than a hundred employees, a considerable number (it is said 45 per cent) are still in family hands. Studies show that these family firms weathered the storms of the financial crises better than other companies because of the members' trust in each other and the sacrifices they were willing to make.

A way to counter mismatch could be to make workers owners of their company (and in so doing create an artificial family). Cooperatives are in vogue again. As an example we could cite the previously-mentioned Amsterdam microbrewer Oedipus, owned by everyone who works there. Together, they are responsible for everything. If someone is cutting corners, others will give them a rap over the knuckles, and this too is very much like the ancestral way of interacting. Likewise, reintroducing the old cooperative banks has been mentioned as a solution to the banking crisis: if it had been their own money, bankers would have thought twice before behaving so recklessly.

In addition, the way offices, workplaces and buildings are designed could help cancel out mismatches. Any resistance to the office landscape notwithstanding, as employees we no longer like sitting in a closed cubicle all day, but instead wish to be part of a larger space with plenty of greenery, which gives ample scope for consultation and gossiping. And, obvious though it may sound, there is such a thing as the 'open door policy' in organisational studies, a recommendation to staff members, managers, supervisors and even CEOs to leave their door open 'in order to encourage openness and transparency with the employees of the company'. The result of an open door policy is an atmosphere of cooperation and mutual respect. And it keeps the air moving, which is good for companies struggling with air quality problems or sick building syndrome. The greener the working environment, the closer it will match our natural past. One of this book's authors' research programmes aims to chart the effects of this (see Chapter 8).

There are occupational health experts who see freelancers as the hunter-gatherer's second cousin twice removed. People who are self-employed pursue a daily quest for a wildebeest or a field with fruits and nuts, while people in full-time employment could be compared to farmers. In the past, our ancestors were all specialised freelancers who were constantly looking for opportunities. We envisage a system in which the number of people working for themselves will only increase. Perhaps there will be a time when scientists will no longer be employed by a particular university, but will be contracted by Oxford at one point in the year, followed perhaps by Leiden, Stanford or Edinburgh at a later date in that year.

There may also be better ways to ensure that working and living are more aligned. Someone who lives in Leeds and works for an accountancy firm in Manchester should be able to exchange with an accountant who lives in Manchester but works in Leeds. Or do away with the time clock and make working hours truly flexible. Workers are able to decide for themselves when they are at their most productive. And yet homeworking has its drawbacks. Working from home may be flexible, but people who chase after wildebeest and fruit at home can become socially isolated, with all the attendant problems. Homeworkers lack the environmental cues to tell them when it might be time to have a break, or go and have lunch with colleagues, for instance. When these cues are missing, studies show, people work much more and for much longer than they would ideally like to, and they are unable to let go of their work. For these types of mismatch the freelance café has been invented, somewhere for people to work and drink coffee at the same time.

As has been said – repetition reinforces the message – people should be more physically active during work. Solutions for this are gyms at work, standing desks, broken-down lifts and company bicycles.

To conclude, we would like to recommend heartily modern types of ancestral barter economy. They can lead to close (local) social networks, although we would to ask you to refrain from using book-swapping websites.

CHAPTER 5

Follow the leader

It was a strange, yet poignant moment just after midnight on 17 July 2014. The few people interested in politics who had stayed up were able to follow what happened, because the press conference was streamed live. For the umpteenth time, the leaders of the twenty-eight different countries that make up the European Union had not managed to reach an agreement. Some top jobs needed filling, but the EU summit convened for this in Brussels, was not able or did not want to make a decision.

At the nocturnal press conference to tell journalists and the people of Europe that a summit had yet again failed miserably was the German Chancellor Angela Merkel. And she found this difficult, as she was flying high at that point, despite the skiing accident that affected her earlier in 2014.

The previous years she had painstakingly built and strengthened her power base in Germany, as well as in Europe. That week, she had successfully expelled a CIA chief from her country for espionage, thus showing her muscles to the Americans and the rest of the world, and living up to her nickname the 'Iron Chancellor'. Polls showed her to be well ahead of her political opponents and allies. Her approval rating was at an all-time high of 77 per cent.

Three days before meeting her fellow leaders, Angela Merkel, or

Mutti, as she is called in her own country, had been cheering with the players of the *Mannschaft,* her national football team, in a heaving, steaming men's dressing room in Brazil, as they had won the World Cup. Merkel's love for the people's game was genuine, she had screamed and danced lustily with her boys and coach Joachim Löw, who had led the German team convincingly to victory in the tournament. For the players, Merkel was their lucky charm: for her they went the extra mile.

And now she was sitting in a room with a few dozen bored journalists who wanted to know why the discussion between government leaders had come to nothing yet again. Europe was facing major problems and the continent's wish to appear effective was being destroyed by the constant indecision. This pained Merkel. In 2011 she had been crowned the unrivalled European leader by the *Washington Post*. 'Using cunning, stubborn gymnastics,' the paper wrote at the time, 'Merkel has bent an entire continent, and perhaps the entire world, to her will.'

That a relatively small woman like Merkel with a gentle baby face was able to get to this position is extraordinary, because she hardly meets the stereotypical image of a charismatic ideal leader who can move people with vision and views that fire the imagination. Merkel looks dull, solid and inconspicuous, all characteristics that have been proven to be a handicap in acquiring an executive position. Merkel is not the dominant alpha female on a monkey rock, she does not have her predecessor Gerhard Schröder's loud mouth or Helmut Kohl's big-headed obstinacy. She is pragmatic to the core, however, and characteristic of her style of leadership is her waiting game. Not for nothing *Time* magazine wrote about her: 'She exerts as much influence by what she doesn't do as what she does.'

As a woman, she is troubled by the prevailing paradox that 'a good leader' and 'a good woman' seem to be mutually exclusive. Ideal leaders are goal-oriented and able, ideal women are empathetic and loving. Whether we like it or not: male and female leaders are treated differently.

And this also happened during the press conference in Brussels, in the early hours of 17 July 2014. Shortly before the journalists were allowed to ask questions, the German TV correspondent Udo van Kampen launched into a birthday song. Angela Merkel had turned sixty at the stroke of midnight. Van Kampen tried to encourage his colleagues to join in with him singing 'Happy birthday to you, liebe Bundeskanzlerin'.

Dear Chancellor... A male leader would never have been serenaded by the assembled journalists like this.

'Perhaps I should have joined in, that might have been better,' Angela Merkel responded fairly dispassionately, whereupon she turned to the order of the night: the immense indecision of the European Union. Later on her birthday the festive feeling acquired a nasty aftertaste when some 2000 kilometres away in the Ukraine, Malaysia Airlines flight MH17 was shot down with 298 people on board, including 196 Dutch citizens – and Europe yearned for leadership.

Alpha animals

What is leadership? Drily stated it is 'the way in which one or more individuals within a group attempt to influence the other individuals within the group to persuade them to achieve a particular common goal'. In the words of the one of the authors of this book: 'Leadership is a process of social influencing, whereby a leader coordinates the activities of one or more followers.' The OED gives the following definition: 'The dignity, office, or position of a leader, esp. of a political party; ability to lead; the position of a group of people leading or influencing others within a given context; the group itself; the action or influence necessary for the direction or organisation of effort in a group undertaking.' Leadership is not a uniquely human trait, because it can be found amongst many animal species that live in social groups, from the honey bee and the ant to the elephant and the ape. Horses have clear leaders, for instance, a leading mare, who heads the herd in search of food and watering holes, and a leading stallion, who acts when danger from outside threatens. Sandra Geisler, owner of the Dutch company Lighthorse, which introduces business managers to horses, explains: 'Amongst horses, leadership is not defined in terms of who is the biggest or the strongest. It is about: who looks after the others best, who is trusted most by all and who is the most instrumental in keeping the group together and healthy. Leadership is therefore not enforced, but bestowed.' This resembles the model of leadership amongst humans, as we will see shortly.

For our primate cousin the gorilla and, to a lesser extent the chimpanzee, by contrast, leadership is most definitely a question of who is the biggest and the strongest. The most imposing individuals do not have the role of leader imposed upon them, they simply claim it. When apes (and other animal species) compete amongst each other for food and sexual partners, the stronger animals come off best at the expense of the weaker specimens. When a weaker animal submits to stronger counterparts, he avoids aggression and increases his chances of survival. In this way, the so-called dominance hierarchy, characteristic for many animal species, reduces violence within the group.

Having said that, we cannot talk of true 'leadership' – according to our definition – amongst gorillas, baboons and chimpanzees. Dominant individuals are not actively engaged in coordinating goals; they really just do what they feel like, and the weaker apes copy them. Big gorillas can rove the forest with impunity as they are strong enough to be able to defend themselves; smaller gorillas follow the bigger ones hoping they will be protected by them when things come to a head. This type of leadership is therefore an offshoot of being dominant.

In our closest genetic relatives, the chimpanzees, we see more true leadership. Dominant behaviour is kept in check to an extent, as the weaker animals are able to work together. The less dominant individuals are able to form social coalitions to make the alpha male's life a misery and to depose him in tandem.

Primatologist Frans de Waal, in his work *Chimpanzee Politics*, has given a beautiful depiction of how these coalitions are formed in practice. In 1982, in the ape colony of Burgers Zoo in Arnhem, the Netherlands, a lingering power struggle came to a head between three males, two of whom conspired together against the most dominant individual. The entire group was dragged into an escalating conflict. The urge for power and reproductive advantage – or rather, sex – led to betrayal, intrigue and a gruesome murder. The book that De Waal wrote about these dramatic clashes and their tragic ending was recommended by the former Republican speaker of the American House of Representatives, Newt Gingrich, to anyone who wanted to enter into politics.

Ancestral leaders

Leadership in humans is a Darwinist mystery as it frequently involves an evolutionary cost. In ancestral times, leaders would be at the front of their forces during battle or while hunting – leading by example – and would run great personal risk. Why would someone want to be a leader during prehistory when he or she might come to a sticky end? Sometimes leadership was accidental, sometimes planned. Becoming the leader might be to acquire status, to play the boss, increase one's sexual chances, or to do what is best for the group. In fact, these are the same motives for the leaders of today, from Macron to Putin, from Merkel to May.

To make a leadership position attractive, a leader had to be rewarded by fellow tribesmen and women. Leaders acquired more prestige and authority than the average tribesperson due to the

contributions they made to the well-being of the group. This translated into greater reproductive success. We call this 'service for
prestige'. Capable warrior leaders, hunt leaders and diplomats do
indeed have more women and children than non-leaders in modern
hunter-gatherer groups such as the Hadza in Tanzania and the
!Kung in Namibia. Biologists regard leadership as a peacock's tail.
Despite all the associated risks of being leader, he exudes: 'Look how
well I lead! In the process I demonstrate that I am (1) strong (2)
dominant (3) unafraid (4) the cleverest and (5) that I care about the
group. In other words: . . . er, ladies, have my babies!'

For ancestral humans, the reward for leadership was not nearly
as big as for some ape colleagues, where power is directly translated
into additional progeny. All gorilla babies are the alpha male's
offspring. Amongst our egalitarian ancestors, leaders were awarded
certain privileges when they had led their people well, and this
enabled them to maintain more women and children, but the
differences between leaders and non-leaders were relatively small in
this respect.

How did the differences in leadership between humans and
anthropoid apes come about? The answer can probably be found in
our ancestral living environment, the savannah. To put it crudely,
we climbed down the trees to start living on the ground. On flat
grazing land there are many dangers, and it was hugely important
for the group to be able to defend itself against bloodthirsty cats and
antagonistic tribes. The group turned out to be the prime survival
mechanism for humans. A strong group helped with survival on the
savannah. Cooperation became the key human word. Cooperation is
at the root of our large human brain.

Cooperation and dominance do not suffer each other particularly
well in practice. How can you cooperate with someone who
dominates you? Cooperating in a group context is more effective
when aggressive and dominant types are ejected or when the group

take them down a peg or two. No one is stronger than the group. A person with dominant tendencies may be left behind or shut out from the group, resulting in certain death. The early tribes had various defence mechanisms at their disposal which behavioural scientists call STOPs: Strategies To Overcome the Powerful.

These adaptations evolved to ensure that tribes were not dominated by individuals who were in pursuit of their own success and in so doing would undermine the strength of the group. Status was exclusively acquired by someone who served the group well and thus deserved leadership. A few muscles and a mildly dominant character obviously came in handy, for example in the fight against other tribes, but true status would only be invested when someone submitted himself to the group.

Followers and those who are followed

In humans, leadership is a little more widely spread than in other animal species. Someone would take the lead on the savannah in the limited arena in which his talent was able to flourish. Leaders were by no means prime-minister types who had to know about everything, from trade and education to agriculture. Leadership was by no means a full-time business, either. A leader was someone with a good plan or good achievements. If Pete suggested going out on a hunt even though Pete had never come home with as much as a Bunyoro rabbit, all the men would carry on lazing under the trees. But if Jack, who was extremely proficient at catching wildebeest, proposed a walk everyone would definitely follow.

The word 'follow' is important here. There were followers and those who were followed. Amongst our ancestors the followers fundamentally created the leader. Someone came forward and the rest chanted in unison: 'You will sort out this problem for us'. Charismatic people with effective actions or ideas attracted attention. Hunters who caught a lot of meat were one up on the others. When there were problems in the group, people listened more to charismatic people. Good hunters or diplomats were (and still are, by existing hunter-gatherers in several places on Earth) called 'Big Men', because they tended to be literally big men (see the work of the anthropologist Marshall Sahlins). Yet these big men were not official leaders, nor was there a well-defined hierarchy. Spread leadership ensured that everyone had their own role in ensuring the preservation of the group.

Western explorers who travelled the world during the eighteenth or nineteenth century found the phenomenon of spread leadership difficult to grasp. They would frequently encounter tribe members in Africa or South America, whom they would ask to take them to their leaders. The tribe members invariably replied: 'We have no leaders. We are all leaders and followers.'

Esteban Lucas Bridges, son of a missionary and later a missionary himself, grew up amongst a group of nomadic hunters, the Ona people, on the southern tip of South America, Tierra del Fuego. In 1948, he published a much-praised account of his stay, with this passage about the Ona people's flexible attitude towards the issue of leadership:

The Ona had no hereditary or elected chiefs, but men of outstanding ability almost always became the unacknowledged leaders of their groups. Yet one man might seem leader today and another man to-morrow, according to whoever was eager to embark upon some enterprise [. . .] A certain scientist visited our part of the world and [. . .] I told him that the Ona had no chieftains, as we understand the word. Seeing that he did not believe me, I summoned Kankoat, who by that time spoke some Spanish. When the visitor repeated this question, Kankoat, too polite to answer in the negative, said: 'Yes, Señor, we the Ona, have many chiefs. The men are all captains and the women are all sailors!'

In prehistory, important group decisions (such as moving on or attacking an enemy) were taken by several tribe members, exactly as is done by present-day hunter-gatherers – like the San, the Hadza, the Inuit. This happened on the basis of consensus. There was a form of democracy, whereby adult men would sit down together in order to decide on the strategy, rather like a parliament. Striving for consensus was possible, as the groups in which our ancestors roamed the savannah were relatively small and were not allowed to disintegrate. Anthropologists believe particular individuals – older men with experience – had greater influence and showed leadership. Their influencing was very subtle. What do you think, and what about you? Once everyone had had their say, there would be a summing up. If there was still no agreement, the group leader could steer everyone towards a decision, but this too was done subtly. Presumably people did not like voting in ancestral times, because there were risks attached to voting. A vote could always lead to a fifty-fifty result, with all the ensuing problems for unity within the group. They preferred to continue talking until consensus had been reached.

The world-renowned Dutch polder model – in which the government, employers and employees come together to take decisions about pay rise and pensions – to some extent resembles decision-making in prehistory. When exactly the term polder model became fashionable, cannot be ascertained with certainty. In print it was (probably) first coined on 10 January 1997, in an article in the Dutch newspaper *NRC Handelsblad*. Internationally, people began to notice the Dutch consensus model and its blessings were lauded throughout the planet, from the White House to Brussels. The Conservative-Labour 'purple' cabinets managed to create many jobs, they had balanced budgets and worked constructively with both employers and unions. The 'Dutch miracle' became an export product that was copied by other countries. In 1997, *Business Week* called the Dutch consultative structure 'the Tulip model'.

Yet the polder model was not invented during the final decades of the past century. People from the provinces of Holland and Zeeland had developed a way to work together to keep their heads above water as far back as the Middle Ages. Literally. At the time everyone simply had to work together to protect the (reclaimed) land against turbulent waters. Farmers, lords, city and country people combined forces, without too much hierarchical bother, to build dykes, reclaim land, and to solve problems through dialogue.

This form of cooperation stemmed from our evolutionary past, from the times when groups were only able to survive by working together. Traces of both this primeval past and collaboration in the Middle Ages are still evident in present Dutch society and politics. The message is: consensus rather than hierarchy and soft, joint measures rather than hard, unilateral decisions. This egalitarian ethos is omnipresent. The Netherlands is one of the best performing welfare states in the world, and there are few places on Earth where the gap between high and low incomes has narrowed to such an extent. More than 90 per cent of the Dutch avail themselves of some social measure or other, and the financial contribution made by the sick to their treatment is exceptionally low. The Netherlands is World Champion in 'Act normally, that's crazy enough'. The day after an Olympic champion has received a medal, he or she will have to join the end of queue in the shops. This is an echo from prehistory.

Followers' brains

In his book *The English*, broadcaster Jeremy Paxman cites an amusing anecdote about leadership in which the bishop of Norfolk says to his successor: 'Welcome to Norfolk. If you want to lead in this part of the world, find out where they are going. And walk in front of them.'

Britain has a far less egalitarian society than for instance Denmark,

Norway or the Netherlands, and yet all these countries demand roughly the same from their leaders. In a worldwide study of sixty-two cultures, the University of Pennsylvania GLOBE (Global Leadership and Organisational Behavior Effects) research programme found a remarkable consistency in the way effective leaders were defined. The most important characteristics are: having integrity and competence, and being visionary, inspirational and self-sacrificial. Throughout the planet ideal leaders are expected to have integrity, be generous, honest, diplomatic, resolute, intelligent, competent and visionary. This prototype corresponds to leaders in ancestral society.

It is clear that this prototype has been shaped by millions of years of evolution. Those in prehistory who followed the wrong leader with the wrong characteristics simply did not leave any offspring. When someone conforms to this prototype, he is a leader – whether he is white or black, lives in the African bush or has a cabinet post in Great Britain, or whether he is male or female. In Asia, an authoritarian leader is esteemed more than in the West. In Anglo-Saxon countries leaders are supposed to exhibit their status by driving in a big car, for instance, whilst in the Netherlands, we consider it perfectly normal (or would like to consider it perfectly normal) if the CEO or prime minister rode a bicycle to work. The leaders of big organisations leave the implementation in the hands of middle management – and that's where things go wrong. So-called remote leading – as opposed to face-to-face leadership as in ancestral times – can lead to great excesses.

Scientific literature makes a distinction between so-called 'transactional' and 'transformational' leaders. A transactional leader is very sure of himself, is alert, keeps a close watch on others and tells his subordinates what to do. His leadership style is hard and business-like and focused on punishment and reward. A transformational leader is charismatic and has inner motivation that appeals to people's imagination. He aims to coach his followers in such a way that they perform above expectation. His leadership style is personal, warm and inspiring.

The difference between transformational and transactional leadership can be illustrated through the leadership styles of Dutch top chefs Jonnie Boer and Sergio Herman, whose cooking has landed them both three Michelin stars in their respective restaurants De Librije and Oud-Sluis. Their cooking philosophies resemble each other, but the men direct their kitchen troops in their own personal way. Jonnie Boer is a transformational leader and a father figure, who shows his chefs how to do things and tries to teach them culinary insight. This is the slow route to learning.

Sergio Herman, on the contrary, is much more of a transactional big-brother figure, who stands amongst his boys and regularly seeks confrontation. His book *Eat, Drink, Sleep* (*Eten, drinking, slapen*) from 2012, an account of cooking at global top level, features a scene describing the day Michelin inspectors visited the restaurant. Herman gave a pep talk to his kitchen staff to make sure they were focused. Whilst he was doing this, he saw a waiter make a mistake. When serving a particular dish, he was meant to put on white gloves (customary in these kind of restaurants), but the chap had chosen to do this in the dining room instead of in the kitchen. Sergio saw this through the kitchen window and flew into a rage. When he returned the man received a mighty dressing down.

'I want you to be razor sharp, man,' Sergio Herman said, where-upon he turned to everyone in the kitchen. 'This goes for all of you.'

Later the waiter revealed that he did not mind having been ticked off in public.

'I've been working here for ages,' he said. 'It was particularly hectic this afternoon because of the Michelin people. Sergio needs someone to point out to the entire team how important it is we don't make any mistakes. I know the situation and am happy to play my part in it.'

This is the higher psychology of leadership.

Women and leadership

There is an infamous tale in TV circles about a female hotshot who held a high position in a production company. A commercial TV-channel was looking for a new programme director and they spotted this woman. She sold her skin dearly, conducted rock-hard negotiations and was able to bag a fabulous deal in the end. When the contract was due to be signed the channel opened a bottle of champagne. After she had put her signature down, the high flyer said: 'Not for me, I'm pregnant.'

The TV executives were stunned.

'You should've told us when we offered you the contract,' one of the men said.

'Would you've offered me the job if you'd known I was pregnant?' the woman riposted.

This anecdote illustrates one of the difficulties women experience concerning leadership. In modern times women are often still not accepted as official leaders, as they do not meet existing stereotypes about leadership. The proposition 'think leadership, think man' is deeply engrained in the human brain and is probably the result of our evolutionary history in which leadership, all said and done, was a physical masculine affair. But it has become an obsolete cue our brain continues to respond to. The glass ceiling is probably a corollary of the Stone Age. Nowadays, leadership is much less physical – even wars are fought from behind a desk – and many women have qualities that would make them particularly suitable to lead in socially complex organisations.

Studies show that women have more empathy, are verbally more adept, socially more sensitive – and much more so than men – are oriented towards maintaining good social relations. Women are also less focused on hierarchy, something that could be an advantage in modern times in which companies are increasingly operating as network organisations, with professionals who need little direct management.

And yet women are still lagging behind when it comes to holding top positions. According to Statistics Netherlands' Emancipation Monitor, the number of women in top jobs in the one hundred biggest Dutch companies rose steadily from 10 per cent in 2011 to 15 per cent in 2013; still a lamentable figure, of course. Science does not perform much better: between 2010 and 2012 the number of female professors increased from 13 to 16 per cent. The Dutch government is a better employer for women than business or science: 28 per cent of top positions were held by a woman in 2014. The not-for-profit sector does best: the proportion of top women is 30 to 35 per cent.

For that matter, the Netherlands' result with regards to female leaders in business is deplorable in an international context. A comparative study by management consultant Grant Thornton from 2014 reveals that the Dutch score was well below the global average. With 37 per cent and 35 per cent women in top jobs, Eastern Europe and Southeast Asia demonstrate that female top managers should be taken seriously. Russia scores even better with 43 per cent, a relic from the former communist ideology. Pitifully low down the list are countries such as India (14 per cent) and the United Arab Emirates (14 per cent). The Netherlands occupies the second-lowest position, with 10 per cent, followed by Japan with 9 per cent. The UK fares better with about 25 per cent of women directors in the FTSE top 250 companies.

Because of our Stone Age follower's instincts, we evidently cannot always see women as suitable leaders. In order to be accepted, women with leadership aspirations occasionally need behave atypically (read: in a masculine way). Margaret Thatcher loved to be called The Iron

Lady, to emphasise that she was a match for her male counterparts. She had speech lessons to lower her voice and went to war against Argentina (over its annexation of the Falklands) to show her strength. The German Chancellor Angela Merkel, who does not have any children, is in her behaviour and appearance masculine rather than typically feminine. It seems that women are only able to achieve leadership when they are not able to have children or give a clear signal they no longer want to have children. Sixty-nine year old Hillary Clinton, a grandmother with a granddaughter, could have ended up the most important person on Earth.

Lord over yeoman

What happened after the introduction of agriculture? Leadership was, in scientific terms, 'formalised' and 'institutionalised'. This meant that someone was allowed to call himself officially 'leader'. Words were introduced for 'leaders', as they did not exist before. And the advent of agriculture ushered in houses, buildings, new codes of behaviour, better weapons, different food. New laws were drafted, accounts and records had to be kept, a need for agreements and contracts sprung up. Ways to formalise agreements were introduced. People learned to read and write.

The enormous population explosion led to unprecedented coordination problems. Who should manage food supplies? How were farmlands watered? Who dug the irrigation systems? Who protected home, hearth and stores? How should property be dealt with? Who inherited houses and country estates when the owners died? A need for a fixed and stable hierarchy, for individuals to fill specific tasks full-time, soon emerged. This mutual hierarchical connection came into being in order to coordinate the increasingly large groups of people. Tasks that had been carried out unpaid by one single individual in ancestral times expanded on such a large scale that they required several people. Complex administrative layers and other matters evolved which, as simple primates, we had never been confronted with.

The transition from leaderless communities to communities with institutionalised leadership happened more or less simultaneously across the entire world. Communities with powerful kingpins had more resources at their disposal than communities without leaders. People who, directed by a powerful leader, dedicated themselves to irrigation projects or the protection of farmlands, would ultimately vanquish the anarchists who did not. Societies with a police force to punish scroungers fared better than lawless ones. Electing leaders and submitting to hierarchy has turned out to be an excellent idea in recent human history, an idea that has led to a surplus in resources, the ability for groups to grow in size and greater prosperity for all. Or at any rate: greater prosperity for the rich. A massive social stratification was the result of these developments; over the centuries, kings, emperors, supervisors and managers surfaced, but also subjects, subordinates and slaves.

Hierarchy was new from an evolutionary viewpoint, but our primate instinct responded predictably: individuals who ended up in positions of power began to abuse their power to favour themselves and their genetic relations. In past times 'the group' would have brought these rogues into line, possibly with aggressive means, but thanks to the new societal constellations they were able to go about their business with impunity. Leadership was no longer used for the well-being of the group, but for personal advantage. And this led to a big contrast between rich and poor, patrician and plebeian, the 'posh' and 'chavs', people in the fast lane and those on the other side of the tracks.

The agrarian societies that had been evolving since the advent of agriculture and in which leaders used the extra resources to enrich themselves (by putting their own family, offspring or followers first when dividing up the granary and supplies) were gradually turned into kleptocracies, or plundering regimes. Subordinates did not have the option to leave their patch of land, so they had to tolerate the behaviour of their leaders. Our research shows that when sub-ordinates have exit options, they do not tolerate the rule of an authoritarian leader for long, and they depart. But if you cannot go away – as in authoritarian regimes like North Korea or in slave-wage Apple factories in China – a leader has free reign to become a dictator. If you do not agree, you may end up in a camp or choose to jump off the factory's highest floor.

Leadership was personal

These days, we still value leadership greatly. Bookshops are full of titles about the subject. How to become a better leader? How to change your leadership style? What can we learn from Winston Churchill or Warren Buffett? Reading newspaper articles about companies, we found that they are remarkably often about the company's CEO. If the results are bad, then he will be to blame; if they are good, he will be praised to the rafters. Studies show that this is not always justified, because there are many reasons why a company might perform well or badly.

Exaggerating the impact of a leader points to an evolutionary inheritance which is also called the 'romance of leadership'. This is an instance of mismatch. In the small groups of our ancestors a good or bad decision could make the difference between life and death, but these days this is much less the case. If the CEO of a company fails, the worst that can happen is that he loses his job.

When we compare leadership in prehistory and leadership in current society, we note a few mismatches. These make us follow the wrong leaders, leaders not functioning as they should and our being unable to kick leaders out when we want to do so. Firstly, we no longer know our leaders personally, as we did in prehistory. At that time, we knew our leaders inside out as far as personality and talent were concerned, and there were barely any secrets. There was no separation between a leader's role and the person. It was pretty clear to everyone who was the best hunter (tip: the person who returned with the largest haul of meat), or who was the most effective negotiator with neighbouring tribes (tip: the person who did not end up with his head on a post).

These days, we may catch a fleeting glimpse of our leaders from a distance during an event or at an election rally, but most people know their leaders solely from TV and other media. This gives a distorted picture. When we meet people in real life we subconsciously take note of dozens of attributes. As little as 7 per cent of human communication consists of the meaning of words actually uttered, the rest is non-verbal and is related to intonation and speaking volume. When we sound someone out, it is the movements, glances, vocal pitch and even odours that provide information about this person. If we solely go by what we see of him or her on television, we are basing our judgement on a flawed image. The appearance of a leader is an exaggerated cue that can put us on the wrong track.

An excellent illustration of this is the American TV debate between the then relatively unknown Democrat J. F. Kennedy and the much lauded Republican Richard Nixon. The debate was broadcast both on radio and on television. Nixon had just been discharged from hospital; he had lost a fair bit of weight and had shaved badly. Kennedy on the other hand was in radiant health and resembled a puppy dog ('I had never seen him so fit', Nixon later commented on that evening). Television viewers saw Nixon sweat and physically he looked weak. They regarded Kennedy as the 'visual winner', whereas radio listeners, basing their judgment solely on what he said, declared Nixon the 'audio winner'. In other words, this shows how important visual representation is when determining who our leader should be.

A big gap has developed between leaders and followers. We do not know what kind of person Dutch Prime Minister Mark Rutte really is, what his sexual orientation is and what he does in his spare time. Officially we obviously do not want to know this anyway, but off the record we do. Seeing Rutte, May, Macron or Trump occasionally in a press conference or on a talk show – or our CEO at the company Christmas party – is not enough. Of course we can create an image of someone through the media, but we do not look this person in the eye and in the final analysis we do not know if this person is truly reliable and competent enough to lead us. We principally base our judgement of someone on a few superficial features: outward appearances such as height, a dominant face, low voice and the ability to be somewhat eloquent. The absence of personal information about someone's functioning is a mismatch. Our ancestors were able to go by hard evidence. Nowadays the relevant cues to judge whether someone will be able to lead effectively is missing.

Hereditary leadership

Kings or lines of succession were unknown to the egalitarian hunter-gatherers. You became a leader on account of your individual qualities, not because you came from a particular womb. Birth will naturally have a played a part in someone's status within the group, as is the case in other primate species, but hereditary succession, as it emerged amongst humans following the advent of agriculture, is an unknown phenomenon in human evolution. The fact that modern democracies like the Netherlands, Denmark, Sweden, and the United Kingdom continue to deploy insalubrious and in-bred families loyally giving birth to sovereign heads of state every few years, runs counter to our ancestral follower's instincts. The royal family is a mismatch.

Some scientists believe that in particular circumstances, hereditary succession can provide stability within a group, but it also creates incessant tension between monarch and subjects. The more power a king assumes, the less his people are inclined to grant it to him. Take the story of Henry VIII of England (1491–1547), who did not exactly increase his popularity during his reign. This had to do with the maniacal way in which he tried to impose his will not only on his people, but also on the rest of Europe. The 'King's great matter' was the divorce of his first wife Catherine of Aragon, who was not able to give him the son the king yearned to have. Catherine had six children, but only one, a girl, survived.

As he was hell-bent on having a son, Henry asked the Pope to annul his marriage so that he could legally wed a woman able to give him a baby with the right male equipment. When the Pope refused, Henry separated the Church of England from the Church of Rome. This was the power he had and the power he abused. Many people in England did not agree with his decision, but uttering this opinion would lead to a single trip to the scaffold. Henry would marry five more women after Catherine, two of whom were beheaded through his tyranny. He died aged fifty-five, weighing an estimated 178 kilos. He was survived by one son (who died at the age of sixteen) and two daughters, Mary I and Elizabeth I, both of whom became queen.

It should be clear that succession to the throne frequently caused major problems in history. When there are several heirs it can lead to civil wars and sometimes even the break-up of a country. Family businesses, too, have to deal with the issue of hereditary leadership. When a firm passes from father to son there is only a limited talent pool to choose from and this has led to quite a few family feuds, in the past as well as nowadays. Perhaps the most well-known example is the case of the brothers Adolf and Rudolf Dassler. In 1920, they followed in the footsteps of their father, shoe giant Christoph Dassler, and began to produce footwear in the Bavarian town of Herzogenaurach. Both men joined the Nazi party and produced boots for the Wehrmacht. Brother Rudolf was captured after World War II by the Allied troops and was soon suspected of having been a member of the SS. The person who is alleged to have leaked this fact was Adolf (his brother, not the Führer). After his release Rudolf and Adi, as he was endearingly called, no longer got on. Rudolf left the shoe factory in 1949 to start his own firm in the same town: Puma. Brother Adolf named the remaining company Adidas. The split between the brothers also split the population of Herzogenaurach.

What is the situation concerning the descendants of modern leaders? In prehistory, leadership was rewarded to a modest extent with reproductive success. Successful leaders were more likely to beget offspring than followers. Does this link between leadership and descendants still exist? This is far from evident. A Canadian study amongst people in different positions in business shows men with high status have more sexual relations than men with a low social status. But it remains to be seen if they have more offspring. After the introduction of agriculture, having many children was a status symbol; truly powerful leaders had a harem. The Moroccan ruler Ismail the Bloodthirsty (1634/45–1727) is alleged to have had more than a thousand children (including seven hundred sons, who waged a bloody war over his succession after he had died).

For modern democratically elected leaders there appears to be no correlation between their position and the number of children they father. In fact, since the industrial revolution, there no longer seems to be a link between power and children in the free world. In present times it is the underclasses who have more children. This means an inverse relationship between power and reproductive success. Our primeval brain says: pursue status and power, you will automatically have more children. Former President Obama, until recently the most powerful man in the world, has just two children. Angela Merkel and Theresa May have no children, like Dutch Prime Minister Mark Rutte and French president Emmanuel Macron.

This seems to point towards a mismatch. But it is also possible that the professional and executive classes, as argued in Chapter 3, focus on increasing the quality of their children rather than the quantity; better to have one child that is extremely successful (and able to have many children him or herself) than a number of children who are abject failures in society.

Complexity

According to the 'Peter Principle' (formulated by organisational scientist Laurence J. Peter), in a hierarchy, every employee is eventually promoted to a position above his level of competence, with all the damaging consequences this entails. This is a mismatch, because hierarchies of this kind did not exist in ancestral times.

Another mismatch is the growing complexity of management tiers, which means that particular positions are held by people who do more harm than good, and on the whole did not take up these positions voluntarily. One of the excesses of this complexity stems from our evolutionary penchant for meetings in which a great deal of talking happens and few decisions are made (as was the way of the ancestral group). And what was decided was frequently a compromise; in big organisations this penchant works against us. It is often said that a camel is a horse designed by committee.

Many governance tiers are filled by executives whose main job is to manage instead of lead. We are talking about managers, process supervisors and so-called middle management. In her book *Meetings? Don't do it!* (*Vergaderen? Niet doen!*) science journalist Ellen de Bruin demonstrates that the kind of language someone uses during meetings is closely related to his or her position within the company. More senior people waffle more and for longer than people who are less high up the ladder. The fact that someone is talking does not necessarily mean he or she has more to say. As a stylistic experiment we have concocted a text using the kind of excessively complex language with which committees and policy groups try to mask their lack of leadership:

> The mission reads as follows: the core-defining production processes within the Ministry of Education, Culture and Science have seen drastic standardisation through the implementation of cross-departmental process renewals, as agreed in the present government's policy priority plan, which has facilitated an efficiency improvement. This moratorium of the flow pilot leads to smoother implementation of the new targets, partly as a result of boosting the theme-focused participating output, whereby we split the targets to be met into statistical production (observation strategy, diversification, quality assurance) and information development (public satisfaction); the objective of the delta plan is therefore to consolidate with force primary observation through a fundamental reassessment, to identify an upsizing for the resonance group which will take shape as an extension frame and standard of the differences between provisional and definitive efficiencies; all this as part of the prioritisation scheme that argues for an improvement in the legibility of policy documents. So please do not suggest that the committee is not taking any action on this front!

Leadership requires vision, guts, and the ancient quality of persuading others to follow you in a particular direction. 'Leading' and 'following' are inextricably related verbs. Organisations, authorities, schools and companies are so complex and so large these days that for managers, the emphasis is now on very different aspects of leadership: maintaining order and ensuring that everything runs as smoothly as possible. These qualities involve few of the traditional leadership qualities. Leaders have become managers, but managers are not necessarily leaders. Which is why citizens, employees, teachers and students are not dashing about for their managers and process supervisors. Managers do not tend to inspire and kindle enthusiasm; they only make sure that everything remains as was decided in the meeting. We have brains that respond to leaders, not to managers!

The haves and have-nots

Our reward system for leaders is a mismatch. We receive money for services, instead of exchanging goods. But exactly how much do we really need and how do we know what is reasonable?

The starting point for barter trade was: fairness. Was a jar of honey obtained with great difficulty equal to a solid T-bone steak or a cubic metre of firewood? Michael Norton from the Harvard Business School interviewed people in forty countries to find out what they thought the average CEO and the average unskilled worker earned, and what they considered to be fair incomes. The ideal ratio between the average wage of a worker and a board member was 1:2 according to Danish people and, according to German and American interviewees, 1:7. These people thought the actual ratio was higher: in Denmark 1:4, in Germany 1:17; and in the US 1:30. The reality was astonishing, however. In Germany the difference between someone on the shop floor and someone in senior management is 1:147 and in the US as much as 1:354! Compare this to Denmark (1:48), the Netherlands (1:51) and the UK (1:84). In the Netherlands, an average CEO earns fifty-one times more money than an unskilled worker. The real pay ratio is miles removed from our ideal dating from prehistory: modest status differences between people from the same group. Big status differences are an evolutionary mismatch with troubling consequences.

We have landed in a Darwinist arms race of rising top salaries. Research by Bloomberg shows that the gap between the top and the shop floor has increased by 1000 per cent since 1950. We derive status from our income, and if you cannot meet this ideal you will go down the social ladder. Is the work of a CEO in the Netherlands really worth fifty-one times more than that of a low-skilled worker? In America, a gap now exists between the 1 per cent of the very wealthiest and the 4 per cent below. Someone who has accumulated the astronomical sum of one billion dollars no longer counts in the League of the Super-Rich who have amassed ten billion dollars. According to the annual *Billionaire Census* there are twelve million millionaires in the world, but only about two thousand billionaires (calculated in dollars). The average billionaire is worth three billion. Globally, these billionaires have a total of 7.3 trillion in their pockets, or 4 per cent of the world's entire wealth. A billionaire's average age is sixty-three; the average number of houses he or she owns is four. One in thirty billionaires own a sports club.

The mismatch is that the haves compare themselves only to other haves and no longer with the have-nots in society, even if they work in the same organisation. There was a huge outcry when international bank ABN AMRO's top tier wanted to award themselves a substantial pay increase while general staff pay had been frozen for years. The argument of the executives was that they compared themselves to the heads of other big international banks and accordingly should not lag behind with regard to reward. They conveniently forgot that the international banking elite was not exactly waiting for their services. As Dutch comedian Youp van 't Hek wrote in his column in *NRC Handelsblad*: 'After years of lousy figures, slashed services, closing branches, with thousands of workers ruthlessly booted out, why not fire off a press release to announce that you have given each other an extra hundred grand or so. Surely you're just getting on with your job?'

Big income and status differences undermine solidarity in society. Studies by the British sociologist Richard Wilkinson (published in *The Spirit Level: Why More Equal Societies Almost Always Do Better*) show that people on average experience less happiness in countries with greater income inequality. The so-called 'Balkenende standard' used by the Dutch authorities is a tool to do something about this. This act, named after the former prime minister Jan Peter Balkenende, prescribes that leaders in positions of public office are not allowed to earn more than 130 percent of the salary of the prime minister of the Netherlands.

Swimming with sharks

Can modern society hold bad leaders personally accountable for their actions and decisions? In ancestral times leaders were simply toppled or knocked off if they took unnecessary risks on behalf of the group. Nowadays leaders are able to take illicit risks with impunity. This has been an issue in the financial world, for instance, where extremely complex systems were invented and where the enormous gap between the top and bottom in the financial hierarchy was abused on a massive scale. The financial crisis of 2008 is one big mismatch with a stiff dose of exaggerated and perverse cues. Speculating with other people's money did not occur in ancestral times, simply because money did not exist. We have no biological adaptations to deal with money and so cannot make head or tail of the complex constructions bankers and investors are forever devising for us. Bonuses for perverse financial conduct (for example giving mortgages to people who cannot afford them) unlocks the worst in executives.

As the Dutch *Guardian* journalist Joris Luyendijk analyses in *Swimming with Sharks* (*Dit kan niet waar zijn*), we escaped by the skin of our teeth a total collapse of our financial system. Specialists (mismatch 1) had come up with financial products that no one else understood, but which did generate a great deal of money, without they themselves or the banks running any risks (mismatch 2) while everyone involved was receiving substantial bonuses (mismatch 3), in organisations so enormous (mismatch 4), that their management (mismatch 5) really did not have a clue what was going on.

The dark triad

What happens if leaders gain too much power? Ever since ancestral times power has been an irresistible cue. But in those days, the might of a leader was kept in check as he was only able to take the lead in his area of expertise and was recompensed with a modest reward for this. But in modern times, power and reward have increased disproportionally. As we saw earlier, an American CEO earns 354 times more than his lowest paid employee. What can this power difference do to someone who has a 'dark triad' personality? In human history, there are hundreds, perhaps thousands of examples of rulers who abused their power and subjugated their people with an iron fist. Psychology talks of dark triad leaders, leaders with a deadly combination of narcissism, Machiavellianism and psychopathy.

Narcissist leaders are very full of themselves and expect others to feel the same. They believe they have exceptional qualities, like to boast, want to be showered with admiration and praise only, and feel an aggressive contempt for people who criticise them. Someone with a narcissistic disorder appears to function normally, but is saddled with an overpowering pattern of superior feelings. A fourteen point list exists for checking whether someone has a narcissistic personality disorder. Useful perhaps, to check leaders in your environment or even yourself.

A narcissist has an insatiable need for (1) admiration, (2) finds it difficult to empathise with others, (3) has an inflated sense of his own importance, and (4) will always overstate his own achievements and talents. (5) He expects to be acknowledged as superior, (6) without having to turn in the attendant performance. (7) He has unbridled and unrealistic fantasies about success, power, and standing. (8) He believes his very special and unique personality can only be understood by himself and by other very special people. (9) Because of this, narcissists only want to associate with people of status. (10) They demand to be treated with extraordinary benevolence, (11) abuse people shamelessly, but do not give any assistance themselves. (12) Few can do well in their books, (13) and a narcissist believes everyone is envious of their achievements. (14) In short, a narcissist tends to have excessive arrogance. If your personality tallies with eight or more of these points, you are very likely to suffer from a narcissistic personality disorder. You are bound to know someone in your circle who answers to this description, and if not, all you need to do is turn on your TV and see 'the Donald' in action.

Machiavellian leaders make a cold and calculating push for their own goals, wherein all means are permitted. They use cunning to mislead everyone around them and get them do their dirty work. An example of typical Machiavellian behaviour is found in *Macbeth*, which like no other play shows how the thirst for power can corrupt. Three witches tell a Scottish general that, one day, he will become king of his country; this prophecy changes his life and even more so his wife. It is she who urges him to murder the reigning king, playing an evil role herself as she tampers with the evidence. In order to ensure that his deed is not exposed, Macbeth takes increasingly more violent actions, until he and his wife overplay their hand and meet their end. Although the play was written in 1606, a good four hundred years ago, it remains as powerful as ever. Many dramatic episodes in international politics resemble the play such as the power struggles in Eastern European countries after the collapse of the Berlin Wall or the recent coup attempts in Thailand and Turkey.

Psychopaths are unpredictable, impulsive and literally asocial in that they feel very little or no empathy for others. Psychopaths know no limits when it concerns violence or sensations from alcohol, sex and drugs. For psychopathy, too, a checklist has been set up, by psychologist Robert D. Hare. Points can be scored when the answer to questions on this list is 'sometimes' (one point) or 'definitely' (two points). Non-criminal non-psychopaths (ordinary people in other words) on average have a score of five points, non-psychopathic criminals around twenty, and people who have been diagnosed as a 'psychopath' score thirty or higher. The list runs as follows (and we suggest you take part for yourself or for people in your circle):

Psychopaths possess an eloquence and superficial charm, have an exaggerated sense of self-worth, are prone to boredom, can lie pathologically, are cunning and manipulative, lack remorse or guilt, have shallow affect (in other words, feel only superficial emotional responses), are callous and lack empathy, have a parasitic lifestyle, poor behavioural controls, promiscuous sexual behaviour, display early behaviour problems, lack realistic, long-term goals, are impulsive, unable to take responsibility for their own actions, have short-term relationships, commit juvenile delinquency, violate their probation and are criminally versatile.

The most sinister leaders are those who have one of these three dark triad personality traits – or in true megalomaniac cases, a deadly combination. Dictators like Stalin, Hitler, Mao Zedong and Saddam Hussain each in their own way knew how to create conditions to make their power felt in the bloodiest way (if you would like to know how you score on the 'dark triad' scale, we recommend *The Dirty Dozen*, a personality test developed by Peter Jonason from the University of West Florida). Obviously such people were born during ancestral times as well, but their power remained limited and it was actively suppressed by the rest of the tribe. Now they have a pretty much open field.

Why are these dark triad leaders so popular? Research by Nicholas Holtzman and Michael Strube from Washington University in St. Louis reveals that people feel attracted to individuals with these dark personality traits. These dark types appear charming at first sight, they make a good impression and the world vibrates around them. They are charismatic, which in early times was a sign of competence. The political ancestral leader was able to inspire other members of the tribe to follow him on the basis of personal qualities, but these tribe members knew from their own experience or stories from others whether someone was in fact reliable. This is more difficult to assess in modern times as we do not know our leaders. An absent cue, in other words.

There are plenty of people – and we all know them – who aspire to leadership and power because of their status and not in order to do something with their talents. These people do not want to exercise their power in the interest of the common good, but in order to benefit themselves and their families, as in ancestral times. However, mechanisms to counter excessive power abuse are easily eliminated or circumnavigated these days. This leads to corruption on all possible levels of governance: the 'darker' a leader, the more he will abuse his position of power, the more his family will be given preferential treatment. An example of this would be the Indonesian dictator who made sure his son was fielded in an important football match instead of another much more capable player.

Leaders who are too powerful often have sons who are completely divorced from the world. Uday Hussain, son of Saddam, was a lascivious, cruel crackpot, who according to *The Guardian* regularly tortured Iraqi footballer players because they had missed a penalty or had played badly (Uday did not know the first thing about football, so the story goes). Bashar al-Assad, son of dictator Hafiz al-Assad, rules with an even stronger iron fist than his father. Marko Milošević, son of the Serbian president Slobodan, and Tommy Suharto, son of the Indonesian dictator Haji Suharto, rivalled their fathers in murderous psychopathy. Likewise Dino Bouterse, the son of Dési Bouterse, the criminal president of Suriname, is a chip off the old block (which landed him a sixteen year American prison sentence). As previously stated, hereditary succession is a mismatch, and the dark variant even more so.

The three most murderous dictators of the previous century were Hitler, Stalin and Mao, who all developed a fatal type of remote leadership. Adolf Hitler was more than happy for his subordinates to sort out things by themselves, only to intervene unexpectedly with excess zeal. Mao developed a system whereby his vassals had to 'think in his spirit' and imagine at all times how Mao himself would act (if a wrong decision had been taken Mao could always plead that his philosophy had been wrongly interpreted). Stalin used vagueness and big contradictions (one of his ways to eliminate adversary X was to sigh that 'it would be terrible if someone were to do something to adversary X').

Leaders for the future

What can we do about mismatches in leadership? We could start off by doing nothing and wait for things to run their course. This means for instance that we continue to choose the wrong leaders. At the moment we choose leaders on the basis of mostly superficial qualities and features, such as charisma, appearance or height, without knowing the actual person behind this façade. Our brain responds to an exaggerated cue and disregards the rest. If the Germans had been able to test Adolf Hitler on his dark triad features, they might not have chosen him as leader. When we do nothing and leave everything as it is, this means that we entrust leaders in large-scale organisations such as the EU and international multinationals with ever greater power. This means in turn even greater regulatory pressure, bureaucracy, corruption, etc.

We could also do some things to select better leaders. First of all, it is important to obtain a proper general impression of someone's talents and personality. Doing this solely on the basis of a CV or job interview is not enough. Finding out how someone functioned in his previous employment, especially by looking at how he led and managed, can be of importance in leadership vacancies. Gathering the opinions of former subordinates could be an option. That way, it will be easier to keep narcissists, psychopaths and Machiavellian freaks out of an organisation.

Leadership should become more personal, because people follow leaders they know. Social media could play a role in this, a way for leaders to be in touch with many followers they do not know personally. Political parties traditionally put a lot of effort into canvassing, going from door to door, to talk to potential voters face to face. This rather old-fashioned way of doing politics tends only to happen opportunistically before elections. And yet our brain is geared towards processing personal information about our representatives and this may be a good practice. One of us has been working for years at the VU University of Amsterdam, and only recently met the new members of the Executive Board. The old EB-members made very little effort to create close relations with members of staff.

Studies also show that if you alert people to existing prejudices – such as 'only tall people or men are suitable leaders' – this information is more easily ignored. This could be useful in, for instance, job interviews. It could also help to counter particular prejudices with regards to women as leaders. As we saw earlier, many modern organisations need leaders who are good communicators, are socially strong and have a great deal of empathy. By placing greater emphasis on these qualities women will rise to the top of their own accord. Research by one of us shows that if an organisation accentuates cooperation, the prototype of the most suitable leader shifts from a man to a woman. The glass ceiling may be a legacy from our evolutionary past, but when the context changes, perhaps the need for a less physical type of leadership will change as well.

To people who aspire to be a leader we have the following tip. In ancestral times individuals acquired added value in small, egalitarian groups by doing something unique for the group. People with the most skills in a particular field, in other words expertise, were followed the most. It is therefore advisable in our present time to develop your own niche to acquire prestige. Based on history, humans favour spread leadership in which we trust experts. Think of warrior leaders, diplomat leaders, referee leaders, scout leaders, manager leaders and teacher leaders. These roles all have to be filled in order to allow a group to function properly, but it goes without saying this cannot be done by one and the same person. Modern organisations would do well to have several types of leader under their roof.

To conclude, it would be advisable to narrow the power gap between leaders and followers as much as possible. To achieve this it is important that every organisation has STOPs (the previously mentioned Strategies To Overcome the Powerful) in place. Subordinates should be able to assess their managers and leaders, for instance. Likewise, leaders should be replaced at regular intervals to discourage abuse of power. Giving staff members a place to have a good gossip about their boss, for example around the water cooler or coffee maker, would also be a good idea. Criticising the hierarchy is another effective STOP, for example via the in-house journal. In dictatorial societies it is the critical media that are the first to be given the axe. Humour and ridicule are good. The medieval king had a court jester who was able to criticise the king's performance without being knocked off. Satire and ridicule are powerful STOPs to counter an excess of power.

Self-mockery also helps. In January 2014, ABN AMRO CEO Gerrit Zalm gave a speech to his staff at the New Year's reception, dressed as Priscilla Zalm, his sister. The ten-minute speech revealed that Priscilla had been running a brothel for years and had been only too pleased to give her brother advice on how to run his bank. The core values of a brothel keeper were the same as those of a banker, Priscilla impressed upon her audience. 'With us, the customer is always at the heart of things,' said Zalm. 'We embrace our customers, look for a connection and enjoy seeing them again.' Nor did Madame have any difficulties with quotas for women. 'Women on top? That's our motto!' ABN AMRO staff roared with laughter, and Zalm ended up consolidating his position.

In short, there are several ways to counter leadership mismatches. Our followers' instinct sometimes wrong-foots us when choosing a leader, but by making our modern environment resemble that of our ancestors a little more, we will be able to make the right choices more often.

CHAPTER 6

The god paradox

On 20 January 1985, Bonnie Lou Nettles died of liver cancer, an event that would lead to the mass suicide of thirty-nine people twelve years later. Bonnie, a nurse born in 1927 whose interest lay mainly in the occult, was the wife of Marshall Herff Applewhite, Jr. (1931–1997), an American singer who saw himself as Jesus Christ's successor. Fortunately he was not alone in this, because Bonnie held a similar view. In psychology, this is a fairly common phenomenon. It is called shared psychosis or *folie à deux*. This delusional disorder can occur in people who are closely connected when the dominant partner transmits his delusional belief and hallucinations to his submissive counterpart.

When Bonnie Nettles died the couple were at the heart of a religious sect – called Heaven's Gate – in which members sought the next level in human evolution. This phase was called *The Next Level*. Marshall and Bonnie had renamed themselves Do and Ti, after the alpha and omega of the solfège musical scale. Do and Ti exercised extreme control over their followers; they had precise sleeping and eating times and prescribed what the members of the sect should eat in the kind of detail that included what vitamin pills to take. Having been asked to do so by Marshall six male followers were castrated in order to get closer to a divine status. It was noted that after the castration they felt giggly.

It all started when Bonnie got cancer and had to have an eye removed. She was nevertheless not afraid of dying, because the doctor treating her was ignorant and her husband was convinced that, together, they would have ever-lasting life. When she did eventually die, Applewhite announced that his wife had not died, 'but had left behind her broken-down vehicle'. He began to refer to Ti as his Heavenly Father and would constantly ask for guidance in leading his 'sect of sects' as he called his sect with a sense of megalomania.

And then, in 1997, Comet Hale-Bopp approached the Earth. C/1995 01 as the rock was officially called, was one of the brightest comets of the past decades and was long visible to the naked eye. Hale-Bopp was often photographed and one of the photos showed a fuzzy smudge which, according to Applewhite, could be none other than the spaceship with Ti on board. She was on her way back to Earth to pick up her husband, children and followers – there was no other explanation. And so Applewhite told his sect members that the big moment to go to The Next Level was finally upon them.

On 24 March 1997 it commenced, in a mansion near San Diego. Members of Heaven's Gate ate a tasty bowl of apple sauce mixed with phenobarbital. The people who did not like apple sauce were allowed to stir the tranquiliser into a bright pudding. The goodies were washed down with a decent balloon of vodka, whereupon all members, dressed in identical black outfits, taped plastic bags around their heads in order to asphyxiate themselves.

This happened in waves. Once one group had died, the remaining participants laid them out on their beds and covered them in a purple cloth. They wore armbands with the text 'Heaven's Gate Away Team', they carried a five-dollar bill and their feet were clad in brand-new black-and-white Nike shoes. By the time the last three members had committed suicide, it was 26 March. In the end twenty-one women and eighteen men died, between the ages of twenty-six and seventy-two.

The tragic death of these thirty-nine sect members had an additional unpleasant sequel for trainer manufacturer Nike. The trainers worn by the Heaven's Gate members had been bought on 1 March 1997 by two sect members in a shop in North County for a sum of 548.45 dollars. The choice had been primarily made for budget reasons: the Nike Decade model was not particularly popular and therefore not very expensive. On 12 April 1997 the satirical TV show *Saturday Night Live* broadcast a fake commercial with images of some of the Heaven's Gate dead in which the Nikes were clearly visible. 'Just Did It', was the accompanying slogan. The shoes that had taken a sect up to heaven. The specific model Nike Decade never recovered.

What is religion?

'It is a truism to say that any definition of religion is likely to be satisfactory only to its author,' the American sociologist John Milton Yinger wrote in 1967. There are dozens of definitions and descriptions. Evolutionary scientists start from the premise that you can only truly talk of religion when five criteria are met. Firstly, there must be an omnipotent Supreme Being. Secondly, this Being is capable of influencing a human being's daily life and it is therefore advisable for adherents to remain on good terms with him or her or whatever it may be. For example, there is a God in heaven who observes and is able to intervene when humankind messes things up. A third hallmark is that followers of the Being organise themselves, there are leaders and a group identity. Four: in order to join the Being's movement, followers must undertake and renounce certain things to show that they are prepared to do something to belong. For example, praying a few times days on a carpet with your head pointing towards Mecca, or abstaining from sex before marriage. And finally, all clubs of the Being's followers have very specific rules of conduct about how to treat each other, and there are taboos:

regulations the faithful need to adhere to under penalty of a divine intervention.

It is important to differentiate between what is a religion and what is not. Having these criteria to hand we can consider why Christianity is a religion, and why being afraid of Friday the thirteenth and other forms of superstition is not. Football, likewise, is not a religion. Football culture has a strong group identity, but no gods (maybe demi-gods), nor do we let it lay down the law. Communism was seen as an alternative faith in the past by some people, but it is an ideology, not a religion, as there are no supernatural aspects attached to it. Similarly Jobs-ism, adhered to by people who believe in the thinking of late Apple guru Steve Jobs, is not a religion (yet), as it does not involve any rules of conduct. If there were scribes who noted Jobs-ism's dos and don'ts, whereupon Jobs' followers began to gather in an association with initiation rituals and taboos ('Thou shalt not buy a Samsung Galaxy'), perhaps then a religion might spring up.

Taboos

Taboos are a determining factor for the success of religions the world over: the more rules, the greater the likelihood the religious club will continue to exist. The American anthropologist Richard Sosis from the University of Connecticut conducted a comparative study of nineteenth-century Utopian communities in the US, some of which were set up along religious lines. He looked at the popularity and viability of some two hundred thousand communities and found that the secular communes fell apart after an average of eight years, whilst the religious communes kept going for twenty-five years on average (even Heaven's Gate lasted a little longer). A religious food taboo differentiates between people who are actively prepared to do or forego something for their faith and people who simply do not think it's worth it. The substance of the taboo is of little consequence.

This is why one faith forbids eating pork, whereas the other objects to beef. Sacrifices mean people feel more involved in the group. By introducing taboos religions are able to hold on to fanatical followers, whilst those who invest less in their faith drop out.

One of the most important psychological mechanisms behind the working of religious taboos is cognitive dissonance: the tension that arises when someone is confronted with two contradictory thoughts. For example: you are a Muslim, but you enjoy having a drink. How do you resolve this internal conflict? You could give up religion, but this involves great costs – exclusion by your religious community (in some countries you can even expect the death penalty). A simpler way to neutralise dissonance is to adjust your attitude towards alcohol. The more effort it takes, the more strongly your sacrifice will make you believe in the rightfulness of the taboo, and so the more you will condemn others who do not observe it. Religious taboos ultimately lead to the faithful becoming more extreme in their behaviour and outlook.

Religion in prehistory

Belief in the supernatural is a global phenomenon which most likely existed in ancestral times. Around one hundred and fifty years ago people began to wonder what the function of religion might be, or might have been. Religion tends to be a costly pastime with many ostensibly pointless investments or sacrifices. Why would you live a celibate or monogamous life or hand over a sizable part of your income if this was not offset by certain advantages? And what are these advantages? How did they arise?

Exactly when and where religion alighted on Earth is something beyond uncovering, but many scientists are nevertheless pondering the question. In his book *The World Until Yesterday* evolutionary biologist Jared Diamond attempts to find out why religion came into being and what functions religion might have. It is unlikely that religion was 'devised' by someone to solve one particular problem in prehistory. It is sometimes asserted that a wise ancestor one day realised that religion would be an easy way to offer solace after the loss of a loved one or to prevent their fellow group members beating each other's brains out. Diamond argues that religion probably emerged as a result of a unique combination of traits in our ancestors, which gradually acquired a new function. This is a common phenomenon in evolution, and not only in people. A bird's feathered wings were probably initially used to keep its body warm and only later for flying.

What are the building blocks for religion and religious thinking? To begin with, our brain has the ability to understand the cause, functioning and meaning of particular processes. This allows us to formulate 'causal explanations' when a particular phenomenon occurs. Say, someone did a dance and it started to rain the next day, putting an end to a lingering drought – our brain would soon make a causal link between dancing and rain. The rain dance was born. The association between a random event with a positive outcome is at the foundation of superstition, a primitive form of believing. The study into superstition using pigeons by American psychologist Burrhus F. Skinner at the end of 1950s is well known. Pigeons in a special box were offered pieces of food at irregular intervals. In time, the pigeons tried to gain control over these intervals by repeating, at times they were not fed, their chance behaviour during the periods food had been delivered. If the pigeon did a little dance when he received food, then it would dance again a little later, in the hope that this would induce another feeding moment.

Religion also emanates from people's ability to read and understand the behaviour of others. Scientists call this the 'hypersensitive agency detection' system. !Kung hunters carefully observe lions who are tucking into their prey. They watch their stomachs to see if they are sated and if it is safe to drive them away. This ability to assign meaning to the behaviour of living creatures in time shifted to the ability to assign meaning to non-living things such as rivers, rocks, the sun, the moon, thunder and lightning (the philosophical outlook on life that stemmed from this is called animism, after the Latin word *anima*, spirit). Gradually, people began to formulate causal predictions about supernatural forces, predictions which on occasion were ever further removed from reality. People also began to see divine clues in stones or clouds which looked like an animal or ances- tor. A contemporary example (2015) of this is the image of Jesus that a Mexican woman saw in a tortilla after she had taken a bite from it (before that Jesus had manifested himself in a taco, potato, pumpkin, hamburger bun, pancake and piece of toast).

Another important function of religion in ancestral times was to allay fear of life's hazards. Our ancestors lived in an environment beset with uncertainties, with dangers around the corner at all times, from predators, snake bites and famine, to fallen trees, floods and violence. Religion was able to give them the strength to cope with these uncertainties. It has been proven that people who pray and read psalms during war situations feel less stress, are less anxious and are plagued less by depression. Following on from this function, religion offers solace for all the harm that has been suffered. We lose loved ones, we fall ill or are threatened with death. Many religions solve this existential problem by simply negating death. We do not die, we only begin to smell a little strangely, while our true being – our soul or spirit – departs for a better place. Dying is not something to be afraid of, but a transition to eternal life, a reunion with family members and friends who died previously with, if we chose the right faith and have lived a decent life, a lecture-hall full of virgins to give full rein to our desires into the bargain.

In addition to these functions Jared Diamond lists a few more which probably did not arise until after the advent of agriculture: a well-oiled organisation of society, the propagation of obedience (as direct envoys of a higher power), codes of conduct towards strangers and justifying war between nations.

Modern religion

Following the introduction of agriculture, religion became more and more central to human life. That meant a mismatch was lying in wait. After the introduction of agriculture, life became harder, the average number of working hours increased, nutrition deteriorated, people ran a greater risk of catching infections and suffering physical wear and tear, and had to co-exist with large groups of strangers. As Jared Diamond puts it: societies with more sorrow require more solace (and therefore religion). Necessity is the teacher of prayer, the Dutch saying goes, and this is still the case in present times. In poor districts, countries and social classes, religion features more strongly than in rich ones.

When in the wake of the agricultural revolution large communities began to spring up, new codes of conduct had to be introduced to allow people to live together. If each time you met a stranger you came to blows, society would be undermined. New manners had to be established, with rules issued by gods or supernatural forces to ensure people complied. The population explosion gave rise to a hierarchy in society, and religion was used to justify differences in status. It also suddenly became attractive to take someone else's territory, behaviour that could easily be justified through religion (the crusades and present-day jihadism, for instance).

The history of religion after the advent of agriculture is largely written in blood. Since tribes in early times were mutually connected through family ties, wars were only waged on a small scale and without the help of divine justification. Hunter-gatherers could try and bash each other's brains in, without gods playing a part. Killing strangers did not provoke a moral dilemma, as there were no kinship ties. Once large communities had come into existence with thousands of non-related fellow citizens, a religious system ensured that people did not try and kill each other during a chance encounter. And yet this same religion led these same peaceable subjects to go on murderous sprees against other people. The commandment to leave each other in peace did not apply to neighbouring religions. Many religions have an unambiguous position: people of other faiths can be slaughtered without compunction, preferably as cruelly as possible. Religion became the justification for wars and mass slaughter.

Religious wars have been taking place since the beginning of recorded history. The First Sacred War raged between 595 and 585 BC. The Delphian Amphictyony declared war on the nearby city of Kirrha (on the Gulf of Corinth), because it had unilaterally decided to levy a toll on everyone who came to consult the oracle of Delphi. The latest Sacred War has been playing out this morning, in many places in the Middle East, Africa, Pakistan, Afghanistan, Indonesia and all other regions where different religious groups fight each other with fire and sword.

The religious brain

Are there any signs these days that we have a brain that is religious? First of all, we find it easy to believe things for which there is no hard evidence. We tell our children the most fanciful stories, teach them to believe in Father Christmas, read them fairy tales and when someone has died we console ourselves with the idea that this person will observe us from a better place.

Moreover, our brain makes constant causal links between events that have nothing to do with each other. An example is our belief in conspiracy theories, in which causal explanations are given for particular phenomena which are often far removed from reality. Sites circulate online adamantly stating that the attacks of 11 September 2001 were the work of an alliance between the Bush family and an influential Jewish lobby, that the Moon landings were staged or that climate change is made up by environmental activists with a secret agenda. People like to believe in conspiracies and pass them on with relish.

Rituals and superstition continue to play an important role in our lives. When Wayne Rooney takes a penalty for England we cross our fingers in the hope he hit will the target. Dutch football legend Johan Cruyff would only play in shirt number 14, because it would bring bad luck if he did not. He would also kick a fresh piece of chewing gum onto the pitch. It if landed in the opposition's half, it would be a good game. Superstition is a means to create order in a chaotic world. Rituals and fixed patterns are a way to keep a grip on things. And they appear to work. Research by the University of Cologne shows that people do indeed perform better when they believe they can steer their own fortune. At the start of a game of minigolf, people were given a set of balls with the message that these had proven to be lucky. The players with the lucky balls played significantly better than the players who had been given 'proven lousy balls'.

Likewise, humans continue to have a profound fascination for spiritual affairs. Here is a thought experiment. Say one of the authors of this book has managed to get hold of a pen holder which used to belong to Charles Darwin himself. He would handle this pen holder with a great deal more respect than a comparable pen holder from the same period. The fact that Darwin himself put his pens in the pen holder gives the object a magic quality, as if Darwin's spirit lives on in it. A study by Yale University looked into this spiritual phenomenon. Test subjects were given the opportunity to buy a jersey worn by people such as Barack Obama or George Clooney. If they were prohibited from selling the jersey, the price dropped a fraction, but if they were told that the jersey had been thoroughly dry-cleaned the price plummeted, while the price of an un-washed jersey was a lot higher. We clearly attach great value to the idea that the jersey contains more of Obama than simply flakes of skin and body odour, so much so that we are prepared to spend ready cash on it. It also works in reverse. People who, when considering buying a house, are told that a notorious criminal has lived there, on average attach less value to it than people who are not told this.

Religion as a mismatch

Evolution has led to the human brain being able to think religiously, and to humans being able to organise themselves into religious communities. During prehistory and possibly soon after the introduction of agriculture, religion had an important function, but is this still the case? Might religious thinking put us on the wrong foot in modern times? Perhaps religion is an exaggerated cue or even a fake one that misguides us into taking the wrong decisions. This seems to be the view of Richard Dawkins in his book *The God Delusion*, in which he starts from the premise that religion is a meme that contaminates the human brain and is harmful to those who get infected. This is not quite our view. Religion probably has an ancient adaptive function, but our question is whether religious thinking continues to be functional in a highly changing environment. Below we cite a number of religious mismatches that have arisen because we make decisions on religious grounds, which harm our reproductive interests in the short or long term.

Celibacy is an example that readily springs to mind. Priests and other religious leaders are expected to live a celibate life, and in so doing voluntarily renounce their chance of reproduction. Another extreme example was discussed at the beginning of the chapter. Members of Heaven's Gate took their own lives because they believed in their salvation in the hereafter. And then there is Jim Jones' infamous sect. There have been many preachers in the past who foretold the end of time and who – when this happy event did not come to pass at the predicted moment – gave the course of history a helping hand. The Japanese poisonous gas sect Aum Shinrikyo, now better known under the name Aleph, is an Asiatic example of this. This sect combines an ideology based on Buddhism, Hinduism and Eastern folk religions. *Aum* comes from an Indian word meaning 'force of creation and destruction' and *Shinrikyō* is Japanese for 'teaching of truth'. The sect gained global notoriety when, in 1995, it conducted an attack on the metro of Tokyo with home-made nerve gas. Thirteen people were killed and more than five thousand were injured.

The most current murderous mismatch is religious suicide terrorism, recently mostly carried out by Jihadists. Because of the allure of the hereafter and peer pressure, there are people who fire off rounds of bullets randomly in public and voluntarily strap a bomb to their bodies only to remove themselves and the enemy – read, people from another religion – from the gene pool with immediate effect. Even more amazing in the West is that the people who blow themselves up are young men and sometimes young women. What counts here is that the prospective suicide terrorist is promised that their descendants are awarded a heroic status and will be supported financially.

Mismatch also arises when religious opinions prevail over scientific knowledge, for example where it concerns health. In 1971 the strict Christian community in the Dutch village of Staphorst was faced with a polio epidemic that affected forty-four children. Five of them would eventually die. The reason for this outbreak was that the orthodox Christians refused to be vaccinated, because this was not part of God's fate. If the Lord wanted you to fall ill, then you would fall ill. Although around 80 per cent of people in Staphorst have now been inoculated against diseases such as polio, even today there are still individuals who, as a matter of principle, do not wish to defy divine providence. Another poignant example is the story of Dutch actress Sylvia Millecam, who was diagnosed with breast cancer in 2000. Inspired by a self-proclaimed occult medium,

Millecam believed she was merely suffering from a bacterial infection. Her 'spiritual physicians' treated her exclusively with alternative therapies, homeopathic remedies, magnetic fields (and paracetamol for the pain), while her tumour had reached a very advanced stage. It was to be the death of her.

Another mismatch is that religions may be effective in organising communities where everyone belongs to the same faith, as in the past, but not all that effective at organising communities in which people of different religions have to live together. Modern society is made up of people with different cultural and religious identities who have to try and live alongside each other peacefully. In the UK and the Netherlands people with dozens of different religions live cheek-by-jowl. There are numerous examples of people in secular societies who have become victims of religious violence. Examples are Dutch filmmaker Theo van Gogh, members of the *Charlie Hebdo* editorial staff or the innocent audience members coming to hear The Eagles of Death Metal at the Paris Bataclan theatre and Ariana Grande at the Manchester Arena.

Despite the fact that there are positive aspects attached to religions (such as offering hope, identity, community spirit and giving life meaning), in modern times religious thinking can lead to all kinds of mismatches causing us to make the wrong decisions in the name of our faith. Take the systemic suppression of women. Hindus are known for not being allowed to tread on an insect for religious reasons, but women used to be expected to immolate themselves together with their husband's body following his death, whether under the influence of drugs or not. In the unlikely event that they refused, they would be subjected to severe moral pressure from priests and family members. Although the practice was outlawed by the British colonial administration in 1830, some cases are still reported. A case of mismatch, for the widow who had herself cremated with her husband might have been able to have offspring or at any rate adopt the role of grandmother (see the grandmother hypothesis in Chapter 2). They were robbed of this opportunity by their faith.

We could discuss many forms of religion, but perhaps the patterns are clear by now. As a reminder, we have created an alphabet of religious mismatches. A is for Al Qaida. B is for Belfast, Burqas. Child abuse, circumcision, crusade. Death threats. Evangelisation, end time. Fasting, fatwah, FGM. Gaza. Honour killings. Iconodulism, incarnation. Jihadism. Klu Klux Klan. Lourdes. Martyrs. Naturopathy. Occultism, oppression of women. Quackery. Polio. Relics, religious wars, Rudolf Steiner. Sati, scientology, sects, self-immolation, snake charmers, suicide terrorists. Theocracy. UFOs. Waco, widow cremation, Wilders. X-files. Yoga. Zealotry.

If you abandon mismatch you have nothing to lose

Who can we turn to in the face of so much misery? What can we do about the Book of Mismatches? If we assume that religion is ingrained in our genes, how can we avert mismatch? One option is doing nothing. Will religion disappear? Maybe, when people in non-religious communities fare better than people in religious communities. For instance, because they make more use of scientific knowledge when preventing or combatting disease. But another option is that if we do nothing, religions will massacre each other or drive each other away until there is just one dominant religion which, *en passant*, will get even with atheists as well. Or perhaps atheists will join forces and religion will ultimately disappear altogether. Our prognosis is that this scenario is not particularly likely. Some places on Earth are becoming more secular, in others the number of believers is mushrooming.

The likelihood that in modern societies religion will gradually be replaced by 'a certain sense of the spiritual' is a little greater. The common Dutch term for this is 'Ietsism', or 'Somethingism': the idea that there is something between Heaven and Earth, but this does not need to be given a name or framed within a religion. *Ietsists* have no desire to congregate in a church, but they do believe in a metaphysical force or that there is an explanation for particular phenomena. Research shows that the idea 'that there must be something' is subscribed to by the majority of Dutch people.

Spirituality and *Ietsism* are phenomena which do not engender violence between sections of the population (there has been no recorded aggression between, for example, reiki healers and tea leaf readers who practise mindful Wicca). Pseudo-religiosity flourishes, with a large following, especially amongst women. The magazine *Happinez* is one of the most successful publications in the Netherlands of the past decades (there is also an English version which appears four times a year). Men who state in personal ads that they are spiritual get more replies. Spirituality is 'religion-lite', it keeps some of the woolly aspects of the confessional religions, but the sharp and orthodox sides have been shaved off.

One of religion's functions is to offer consolation and comfort. This role has largely been taken over by good health care and strong social networks, but science has not entirely been able to offer the solace of a hereafter yet. Keats wrote that science is 'dull, cold, sombre and arrogant' and is not able to offer people solace. Richard Dawkins countered this in his book *Unweaving the Rainbow* when he wrote: 'Isn't it sad to go to your grave without ever wondering why you were born? Who, with such a thought, would not spring from bed, eager to resume discovering the world and rejoicing to be part of it?' Science as non-religious source of comfort.

Now that God, in many cases, is no longer there to look down and keep an eye on us and make us give account when we report to Him at the end of our lives, this role has been taken over by other phenomena. In the past an omniscient God saw everything; now we have a secular society with laws, regulations, CCTV, speed cameras, social media, public scandals and Breitbart. Omniscient God has morphed seamlessly into an omniscient Big Brother, who scans our data and exposes us when we are so stupid as to walk into a camera's field of vision. In ancestral times reputation damage was a big problem, and these days, too, people run a great risk of catching a social contamination when they do not submit themselves to the laws of modern media (see also Chapter 9). The question for the future is: does Big Brother render God superfluous?

Religion offers people comfort for the idea that one day they will no longer be. Perhaps religion will disappear once our fear of death disappears. Our brain is not geared to reaching the age of ninety or a hundred. When in sound mind and body people often say they would like to live to at least a hundred and five, but years before that age this desire gradually drops away. If you are reconciled to an approaching death, you do not want to have anything to do with a scornful God to whom you need to give account, let alone with the idea of heaven.

The conclusion is that religious thinking can give cause to mismatch in present times. Only when there are alternatives that fill the same human primordial need as religion – comfort, giving meaning to life, hope, punishment – will they, in time, disappear. While this has not yet happened, humans will not soon relinquish their faith.

CHAPTER 7

War, what is it good for?

At half past six on a Tuesday morning the alarm clock goes off in a bedroom in Roswell, New Mexico, where twenty-eight-year-old Tiffany McGregor had been sleeping peacefully up till now. She sighs as she gets up to take a shower. While she is doing this her husband, a pilot called Aiden McGregor, comes home from his night shift. He opens the back door to their detached house and enters the kitchen in a buoyant mood. Then it happens: lying in wait for him are two vicious looking attackers.

'Hands up!' one calls, threateningly pointing his pistol at him. The other intruder stands by, his firearm loose in his hand. Aiden starts. It's a cowardly ambush. They are armed and he is not.

'Don't shoot!' he calls, somewhat panic-stricken. He raises his hands in fear.

'Give me one good reason,' says one of them.

Aiden thinks.

'Because I'm your father?'

'Anyone could say that,' a six-year old boy shouts, and fires his toy pistol repeatedly. A toddler aged four also begins to shoot. Aiden collapses onto the sofa, his hand covering the bleeding hole in his chest. He rolls onto the floor. His children roar with laughter: dad is bleeding to death on the carpet!

When Tiffany enters the room, she sighs.

'We've killed dad,' her eldest son yells. She heaves another sigh.

Ten minutes later the boys are playing a war game on their computer, as Aiden buries himself in the morning paper which is full of reports about global conflicts. The game his sons are playing is a first-person shooter, an action game in which the player views the world from a 'first-person perspective'. Aiden has no problem with this, but after a few minutes his wife thinks enough is enough.

'Just leave them to it,' Aiden says. 'It's just a game.'

'How was work?' she asks, placing a breakfast of hot pasta with ham in front of him.

'Busy', he says, and turns a page of his newspaper. Atrocities in Syria and Nigeria; attacks in Iraq and Afghanistan.

Aiden did indeed have a busy night with his squadron, stationed at an air base half an hour's drive away from Roswell. They were one man short due to flu, so the remaining pilots were twice as frantic in their container. His shift had started with the usual observations of Afghan Taliban freedom fighters. Aiden and his colleagues monitored possible targets and suspects on twelve computer screens, ventilators buzzing all the while. Joystick in hand and headphones on his head for radio contact with people he did not know personally, Aiden studied the screens. At times, he felt like a voyeur. He had seen a bearded man in a robe who, overcome by diarrhoea, had been squatting for thirty minutes in a remote field whilst feverishly trying to shoo away playing children. He saw how a boy in a black waistcoat was being reprimanded by adult men in turbans. A woman was preparing a meal near her stone house. Everything was being registered.

Once they had fixed on the targets, one of his squadron leaders had issued the command to use the MQ-9 Reaper, an unmanned war plane able to transport around 1700 kilos of bombs and rockets. The aircraft took off and soon the drone had approached 'Target X'. It had been Aiden who steered the Reaper to the location of the hit.

Together with a colleague he made sure they had locked onto the right person on their computer screen, and fired. Sixteen seconds later a 100,000 dollar AGM-114M Hellfire rocket saw to it that one particular Taliban fighter would never have diarrhoea again. Aiden and his immediate colleague high-fived each other. When, an hour later, Aiden left the airless container and stepped into the fresh morning air, the sun was up. It promised to be a beautiful day. Driving back home in his Chevrolet Impala Aiden looked forward to his pasta breakfast and his newspaper.

Primitive hooligans

At the time of writing, thirty-one official wars are being fought globally. This includes the Casamance conflict in Senegal, the skirmishes in Balochistan and the independence intifada in the Sahara. Many of these conflicts are unfamiliar to us – both in terms of the regions in which they are being fought, the people and the causes.

The list of battles from the past is even longer, almost inexhaustible; hundreds of wars were waged that have vanished from our collective memory. Who apart from historians knows anything about the Lelantine War between 710 and 650 BC? The War of the Eight Princes (291-306)? Or the Dutch-Hanseatic War of circa 1440?

War – one group going to battle against another – is of all times. Historians believe the actual Trojan War took place as early as during the thirteenth or twelfth century BC, but there are many indications that well before that large-scale conflicts were being fought. Excavations in Egypt, Germany and America have uncovered mass graves from prehistoric times that are indicative of human violence. The bashed skulls and bone fractures suggest injuries from axes and sharp arrowheads. The mass graves denote organised violence.

Wherever people lived in proximity, evidence has been found of manslaughter and murder. Archaeological research shows that in 90 to 95 per cent of societies, traces of warring have been found. Peaceful societies are an exception in human history. But what is clear is that, in ancestral times, war did not involve two armies facing each other to wipe each other out. Any organised violence was above all 'cowardly'. Our ancestors were good at surprise attacks, which anthropologists also refer to as raiding and ambush killings. This might involve a group of men invading a camp at night to kill one or several victims, then retreating in silence (so-called raiding). Or laying an ambush at a river. When someone from the other tribe came by, this person was either killed (in the case of a man), or kidnapped (a young woman). These techniques of raiding and ambushing are as old as humanity itself. Worse still: probably much older.

Science sees 'coalition aggression', or organised violence, as the foundation of human warfare. Battles with many hundreds of thousands of fatalities, trench wars and massacres have their deep origin in our social ability to form coalitions of two against one. In the previously cited *Chimpanzee Politics* (1982), Frans de Waal showed how chimpanzee males from the lower orders form mutual alliances to deprive higher-placed males of power or even to kill them. Several cases are known in which the beta and gamma males conspire against the alpha to dethrone him, often with the consent of the group's females. This behaviour also occurs amongst other species, from lions, hyenas and dolphins to other primates. The ability to form coalitions, paradoxically, underlies our propensity for organised violence. Or if we want to refine it: our social and sometimes altruistic instincts enable us to wage war. Social behaviour and violence go hand in hand in human evolution.

In chimpanzees, the acts of violence not only take place between fellow group members, they also have a territorial quality. Groups of chimpanzees live in their fixed habitats, which sometimes border the habitats of other chimpanzees. Biologists have observed different groups defend their territory against intruders who come and steal their food. In order to stake out their area, chimpanzees form 'border patrols' to guard their boundaries. Border control is not typically human.

At times, groups of chimpanzee males venture forth to look for undesirable strangers. And when they do come across one in their territory, the heat is on. Primatologists have recorded video images of groups of chimpanzees attacking an unfamiliar male. Screeching, they bite and tug at limbs and sexual organs until little is left of the intruder. These are brutal murders. From a human moral perspective, the violence used by apes could be called distinctly cowardly. The chimpanzees only go on the offensive when they are in a clear majority (for example five to one) and a victory is pretty much guaranteed. They limit the danger to themselves to the absolute minimum. This cowardly behaviour is something we see regularly in the media on CCTV images from big cities, usually around bars and clubs. Groups of young men beat and kick a motionless victim lying on the ground. A wave of moral indignation sweeps the country – yet another victim of senseless violence – but nothing apelike is alien to us.

In chimpanzees there is an evolutionary objective behind killing neighbouring counterparts. Research shows that groups of males systematically annihilate males from other groups. A desired outcome is that this increasingly weakens their neighbours. Science calls this the imbalance of power effect: killing the nearby group's males in particular can create a power gap. When this gap becomes too big, the females of the weaker group will end up migrating to their stronger neighbours. As they have a larger territory they are better able to gather food and everyone within the group benefits: males, females and the young.

In the end, cowardly murders of individuals from other groups pay out in an evolutionary advantage for the aggressor. The chimpanzees' violence happens without rational consultation; the apes have no generals or field marshals who map out a strategy for them. They have no armaments and they do not have to abide by the

Geneva Convention. And they do run a minimal risk of getting injured.

To the chimpanzee, who by and large lives in one fixed place, his territory is much more important to him than to his fellow-primate, the human. Our ancestors, the hunter-gatherers, migrated several times a year along centuries-old routes across the savannah, sharing their living environments in the process. They would often have testy relationships with neighbouring tribes, and agreements had to be made about access to watering holes and other natural resources. This was obviously a breeding ground for conflicts.

Violence was prevalent in ancestral times. In *War Before Civilization* archaeologist Lawrence H. Keeley describes how primitive humans used more or less the same tactics as chimpanzees. Raiding and ambush killings were the order of the day. People did not shrink from surprise attacks on their neighbours and during festive peace talks, adversaries might be ambushed. All in all the death rate amongst adult men as a result of organised violence in prehistory is estimated to be 25 per cent. Within this overall figure, big differences between traditional societies can be discerned. Amongst a people called the Jivaro (in Puerto Rico) the figures is 60 per cent, but amongst the Gebusi (in Papua New Guinea) there were just 7 per cent of male war casualties. By comparison: in Europe and the US, the percentage of men dying as a result of warfare during the twenty-first century is less than 1 per cent. Evolutionary psychologist Steven Pinker argued in his book, *The Better Angels of our Nature: Why Violence Has Declined,* that this reduction in violence is one of the most significant achievements of modern society. (More about this under the heading 'Peace for our time'.)

The male warrior effect

Three principal motives can be found for organised violence in ancestral times: status, sex and salary (the famous three Ss in which

212

salary was obviously not paid out in money or property, because they did not exist yet, but in social capital, for example friendships or exchange partners). When a tribe's men managed to defeat their neighbours, the warriors who had contributed most to the victory were awarded a higher status than the lily-livered individuals who had been looking on from the sidelines. The more enemies a man had killed, the greater his standing, which, in turn, was good for his reproductive success. Or in everyday language: for how many children he had. Science calls this the male warrior effect, a deeply rooted tendency in men to increase their reproductive success by taking part in organised violence.

The anthropologist Napoleon Chagnon has spent years conducting field-research amongst the Yanomami, a bellicose hunter-gatherer tribe in the Amazon rainforest. The highest honour men can earn is awarded for killing members of other Yanomami tribes. These men even have a special name, the *unokai*. Chagnon studied how many women and children the *unokai* had compared to other adult males in the group and found a clear distinction: the more warlike the man, the more sex and offspring.

Waging war was also a means to rouse sexual curiosity in women. Attacking other groups and defending one's living environment can be seen as comparable to a peacock's tail. Where a male peacock tries to impress females with feather splendour which is as imposing as it is affected and costly, men in prehistory tried to impress women with their warmongering expertise. By entering into conflicts with competing peoples as a group, men were able to acquire a larger, more fertile living environment, with the attendant chance to reproduce more successfully than their less belligerent counterparts. This probably meant that sexual selection occurred on war-waging traits amongst men.

In addition, waging war is what biologists call a 'costly signal' (just like smoking, which we referred to earlier). The battle is often dangerous and requires a lot of strength and energy. The winners show that they are able to withstand these hardships and are therefore in possession of a decent physique and good genes. They exude the ability to produce healthy offspring and to protect and maintain their family. Recently, one of us conducted research into the question of whether waging war does indeed increase a man's sexual attractiveness. This study was done at universities in the UK and the Netherlands. Female students were asked to read scenarios about a fictional soldier and to indicate whether they found him sexually attractive. Some women read a story about a soldier called John. In the first scenario John had stayed in the UK or in the Netherlands, in the second he had fought in a mission in Afghanistan and in the third scenario he had additionally been awarded a medal for bravery. What turned out to be the case? John was considered the most attractive when he had fought and acquired the status of hero, whereas there was no difference in the outcome between the first two scenarios. A follow-on study showed that John was considered more attractive if he had been awarded a medal for an act of bravery in war than if he had been awarded a medal for having executed a heroic role during a natural disaster. These results are striking, as soldiers in the Netherlands and the UK do not enjoy a particularly high status, judging purely by their training or income. And yet soldiers – and especially war heroes – are desired by women, our research shows. This applied to both short-term and long-term relationships.

Thus, the increased chance of status and sex appears to offer men significant evolutionary advantages that amply make up for the disadvantages. That is the foundation of the male warrior phenomenon. In *A Theory of Warfare*, evolutionary psychologists Tooby and Cosmides demonstrate how this works using game theory. When, as a warrior, you have a one in ten chance that you will die by fighting doubly hard in battle, but the chance of extra offspring increases by 20 per cent, this produces an evolutionary advantage. It is in your genetic interest, after all, to join in the battle, provided that, for the average warrior, the benefits outweigh the costs.

According to Tooby and Cosmides a 'veil of ignorance' must hang over who will live or die in battle (the German language has an equally splendid version of this term: 'Schleier des Nichtwissens'). Game theory makes clear that you will not join in the fight when your chance of dying is 100 per cent. In order to be able to account for suicide missions during times of war, we should not be looking at biology, but at culture and religion (as discussed in Chapter 6). Game theory also demonstrates that you are best off as a member of a large coalition. When one group decides to send two warriors to battle, while the neighbouring group sends five or six, the second group is bound to win, and each individual within the group is less likely to be killed or injured. A game theory analysis of war shows that early in history there must have been an arms race between groups in order to grow and to be able to mobilise as many warriors with as much weaponry as possible, because 'size matters!' The army parades we see on TV in Russia, China or North Korea are a modern manifestation of this arms race. With it, the leaders give off the signal 'Look how strong we are . . . so don't fuck with us!'

Finally, game theory also explains why if you fight extra hard, you can look forward to an extra-large reward. This was recently demonstrated in a study into the Turkana, a nomadic people in eastern Africa who periodically set out to steal cattle from other tribes. The researchers asked Turkana men and women to assess scenarios of men who did not take part in such raids, because they lacked courage or were physically incapable. The men lacking in courage were judged less positively than men who were unfit, and they were punished more as well.

So for men there are clear evolutionary advantages to waging war. The qualities needed to win a war are physical (fighting aggressively), as well as social (showing courage) and political (ability to form coalitions). This male warrior effect presupposes that an entire range of male qualities, both positive and negative, have come into being as a result of a long history of waging war. We even like to speculate that psychopathy – the inability to respond with empathy to others' suffering, a phenomenon particularly prevalent in men – has not been selected away as it offered some advantages on the battlefield. A psychopathic warrior can reach great heights in battle victories, so psychopaths are tolerated in society in peacetime.

War and sex are also directly linked in human evolutionary history. Since time immemorial warfare has provoked excesses such as rape and bride kidnapping. In *A Natural History of Rape* (see Chapter 3), biologists Randy Thornhill and anthropologist Craig T. Palmer explain that rape is a fairly common form of procreation amongst many animal species. In humans, rape has been socially very undesirable and extremely reprehensible behaviour for as long as we can remember. As we saw, ancestral rapists could expect public shame, exclusion and often physical punishment within their own group. But in war situations, norms and values are different. Without incurring reputational damage, men are able to claim the enemy's women in an attempt to spread their genes, without facing reprisals. This resembles the evolutionary 'sneaky fucker strategy' (which we encountered earlier): if you can get away with it, have a go. War gives men opportunities to have offspring without running too many risks of being avenged.

Another aspect of waging war that points to an ancient relationship with sex is the role of the penis. *Flawed Giant*, a biography of American president Lyndon B. Johnson, contains an interesting story about an informal chat between Johnson and critical journalists. They asked him why the Americans were staying in Vietnam and simply did not admit defeat. Johnson lost his patience. 'Do you really want to know?' he asked, when his listeners continued to interrogate him too harshly. At which point the president undid his fly, got his penis out and declared solemnly: 'This is why.'

Lyndon B. Johnson exposed himself and in so doing manifested: 'Look how resolute we are, we have not been beaten yet!' Typical leader, Johnson. Even now we are still looking for the kind of leaders people would have followed in ancestral times (see also Chapter 5). We have a deep-seated natural preference for physically strong leaders who avail themselves of brawny language.

The connection between penises and warfare goes back into deep human history. A while ago, a note from the pharaoh Merenptah (1224–1214 BC) was found during excavations. In it, he related how big his victory had been over his Libyan enemies. Following a grim battle with a Libyan army, the pharaoh was said to have been handed victory spoils comprising 13,320 chopped-off penises (six from Libyan generals, 6,359 from Libyan soldiers and the remainder from foreign mercenaries). The symbolic value of the chopped-off penis lies in the message that the enemy will no longer be able to have children and therefore not be able to bring offspring into the world who, in future, would set upon your tribe. Although no archaeological evidence has been found for such behaviour – a penis does not fossilise – it seems plausible that these excesses occurred in early tribal combat.

The tribal brain

For many men, waging war was and is a life-changing experience. Much has been written about the romantic nature of warriorhood. The American philosopher Jesse Glenn Gray fought in World War II and wrote about his experiences as a member of a platoon in his book *The Warriors:* 'Many war veterans who are honest will admit [. . .] that the experience of communal effort in battle [. . .] has been a high point in their lives. Despite the horrors, the weariness, the grime, and the hatred, participation with others in the chances of battle had its unforgettable side, which they would not want to have missed.'

War is physically addictive because of the huge adrenaline rush. Waging war produces deep, lasting social bonds between warriors, who experience their group as a family. In interviews, fifty years after World War II, soldiers related that the band of brothers forged during the war was the best thing to have happened to them. This is how deep the impact of going to battle is. Researchers at the University of Amsterdam revealed this 'band of brothers' effect in a hormone study. They were primarily interested in the bonding hormone oxytocin. This hormone is released during intimate emotional experiences, such as breastfeeding a child or making love. But it is clearly also responsible for enhanced group bonding. In the study, in which only men took part, researchers placed a snuff of oxytocin or a placebo in the noses of test subjects. They then made them play a game in which they competed as a group against another one. The result? The oxytocin group worked better together, they trusted each other more and the members of the other group less. The researchers concluded that oxytocin had a 'tend and defend' effect.

Warfare separates the men from the boys: only true heroes dare to go through the dust. There are warriors who become paralysed with stress and abdicate themselves from hostilities. This behaviour is not appreciated by the group. In science terminology this is called a classic free-rider problem: what to do with men who experience the benefits of warfare (higher status, greater chance of offspring), but meanwhile refuse battle? In order to correct this behaviour, soldiers are subject to a system of norms, rewards and humiliations, mechanisms that continue to operate amongst nomadic people like the Turkana we encountered earlier. In modern warfare this behaviour applies just as strongly. The fact that under modern military law deserters can be shot forthwith, stems from an ancient instinct to keep the group strong and punish cowards. The disgrace of being regarded a deserter has such a profound effect that there have been several court cases in the UK in which family members of deserters during World War II wanted their loved ones to be rehabilitated. This can also explain the resistance to LGBT people and women in the armed forces: love relationships between soldiers might happen at the expense of unity within a group.

In other words, there are numerous examples of how our brain, and especially the male brain, has been shaped by a long history of warfare, the traces of which are still visible today. We continue to have a, chiefly male, fascination for warfare. News reports invariably open with accounts about hostilities when somewhere in the world a conflict breaks out. We watch war movies, read books about war and commemorate wars from the recent past. Our entire culture is imbued with images and metaphors related to war, whether in sport ('let battle commence') or politics ('we beat them!'). As children, we play many war games, as adults, fire paintballs until we see red, green and purple; and simulate violent situations for hours on end on our computers. Warfare is ingrained in the human (read: male) psyche.

In present times traces of our tribal brain can still be found in men especially, who continue to be tribalistic. Men, much more so than women, support a sports club in a way that goes far deeper than pursuing a hobby. Likewise, men will be much more motivated to operate as a group and to defend their group against other groups (for example when going out in the evening or onto the football pitch).

These gender differences can be illustrated neatly in a study. We asked men and women to choose their favourite colour and to explain why they chose it. A greater percentage of men (22 per cent) than women (8 per cent) chose a colour for tribal reasons ('I choose red because it is the colour of my favourite football club', 'I choose white because it is the colour of my religion'). Women largely referred to nature or fashion ('I choose red because it looks good on me', 'I choose blue because it is the colour of the sky'). In another study, one of us made men and women play a game in which they were able to keep money for themselves or invest it in a group fund.

On average, men were more inclined to keep the money for themselves than women, unless we told them that their group had to compete against other groups. From that point onwards they began to invest more in the group fund.

The question might be asked whether the status-enhancing effects of fighting wars in prehistory also apply in modern times. One of us has conducted research amongst American Veterans, whereby we looked at the difference between those who had been awarded high honours and regular veterans. The proposition was that a soldier with a medal on their chest had more children than the inglorious one. This proved to be the case. The decorated servicemen had 3.2 children on average and the regular veterans 2.7. Men who have displayed courage on the battle field are evolutionarily one up. When the results of this study were reported at length in the media, one of us received an email from Gijs Tuinman, the youngest Dutch soldier to be awarded the Military William Order, with a question about what inspired us to research this topic. Our reply: evolutionary psychology presents many interesting hypotheses.

Someone who has behaved with valour on the battlefield in his younger years can, as time goes by, increasingly live on his past. Generals themselves no longer take part in battle, for instance, but they do enjoy a high status and a great deal of respect. Societies, men and women, are forever looking for people who have special qualities, and wars lend themselves pre-eminently to displaying these qualities. It is not for nothing that generals who fought in World War II later became presidents or prominent politicians (Churchill, De Gaulle, Eisenhower). On the battlefield, they had won their spurs. Again an example of a costly signal (a peacock's tail): people who distinguish themselves by doing something for the group show that they have moral, physical and mental capabilities to serve the common good.

This sacrifice is appreciated more than the efforts of politicians, for instance, and, regrettably for us, scientists or writers. The question as to how heroic someone has been in his (or her) military career is still crucial in American politics especially. During the election battle between George W. Bush and John Kerry, their individual military histories surfaced. Bush had ducked out of military service, but Kerry produced all kinds of tales about his heroic actions during the Vietnam War. Major doubts were cast over this and the American right-wing press tried all manner of tactics to knock the bottom out of his stories.

Many American politicians have a war record. A stirring example is the American politician Tammy Duckworth, who was born in Thailand in 1968. As an American army helicopter pilot she lost both her legs and part of her right arm in Iraq. Her valour and sacrifice made her a reliable candidate par excellence to make a bid for a political position. She became the first Asian American woman in Illinois to be elected to Congress and the first disabled woman in the House of Representatives.

Women and war

What is the situation regarding women in war? There is little scientific knowledge as yet. But we can speculate. To begin with, women have far less of an evolutionary interest in fighting than men (just as female peacocks have no evolutionary interest in growing such a ridiculously costly tail). What they are able to do is to spur on men to fight for extra territory and food. Women can be calculating and switch loyalties when their group is tasting defeat. The American psychologist Shelley Taylor called this the 'tend-and-befriend response', the female response to threats and stress. Whereas men usually display fight-or-flight responses, women concentrate on the protection of their brood (tending) before anything else and join a partner or social group who can offer this protection (befriending). These might be the male warriors from their original group, but also from the enemy's group if they are winning the battle.

Women may also have developed psychological adaptations to prevent war being waged, or to end a war. Because women by and large have more to lose than men (as a war's evolutionary costs such as rape and the risks to offspring are greater for them), we might expect that women are more inclined to keep the peace. Research shows that women are more negative about the deployment of violence as a solution to conflicts. They are also less likely to go on the offensive in laboratory-simulated war situations. In addition, our own research shows that, in time of peace, we prefer female leaders – either actual women or men with a feminine appearance. It is perhaps not for nothing that the most recent American Secretaries of State included women, Madeleine Albright, Condoleezza Rice, and Hillary Clinton. We expect women, more than men, to be focused on maintaining peace.

Finally, there must have been circumstances in our evolution in which women exacerbated warfare. The best-known (fictive) story is that of Helen of Troy. Research by one of us into the sexual preference of women for war heroes signifies that women are fundamental to war. Sexual selection is the explanation for this. If women did not fall for war heroes, but for bird watchers, many men would spend large amounts of time with binoculars in the great outdoors. Statues would be erected for the best twitchers, and these people would be cheered on by a wild crowd, including countless young, attractive women, before setting out on another expedition.

Male Apache Native Americans were known only to venture out to loot and pillage from neighbours when women complained of lack of food. Women might also have played an active fighting role when they were forced to do so by circumstances. In the event that the tribe had too few male warriors (because they had been killed in earlier fights) women may well have actively taken part in battle in order to compensate for the lack of fighters. Throughout history, this has happened in various places around the world. A contemporary example is the state of Israel, which has been confronted with such superior numbers of enemies that women play an operational role in the army. Like men, women in Israel have to do national service. When there was a large shortage of men during the War of Independence in 1948, women took part in combat. Figures from the Israeli army show that in 2002 women comprised 33 per cent of the lower officer ranks, and 21 per cent were captains and majors. Conversely, they comprised just 3 per cent of the higher ranks, even though, in 2011, Israel appointed its first female Major General, Orna Barbivai.

Evolution plays a strange game with women. Sexual selection has made women fall for male warriors who might form a threat to them during times of war. The enormous scale of warfare has led to the increasing anonymity of soldiers and, with it, an increase in lawlessness and at times sexual violence on a massive scale. In 2008, the film *Anonyma – Eine Frau in Berlin* (*The Downfall of Berlin Anonyma*), directed by Max Fäberböck, was released. It was based on the diaries of an anonymous German woman, who recorded what happened to her and other German women during the months following the Russian liberation of Berlin: a massive wave of violent rape during a chaotic and extremely uncertain time.

The sexual assaults were more than barbarian, the American journalist Cornelius Ryan chronicled in *The Last Battle*, a narrative about ordinary citizens and soldiers from both sides. Ryan recounted group rapes during which many women died and soldiers who forced their way into maternity wards in a drunken haze to rape women who had recently given birth or even some who were heavily pregnant.

In her 1992 documentary *Befreier und Befreite* (*Liberators Take Liberties*) filmmaker Helke Sander made an attempt to work out how many women were raped during the Russian advance. She did this using medical, abortion and birth records and verbal accounts. Sander came to a cautious estimate of 1.9 million Russian rapes in the whole of Germany and in excess of a hundred thousand in Berlin.

For a long time this subject was taboo in Germany, because many German women – to prevent worse – entered into relationships with Russian soldiers as a survival mechanism, often officers (the female tend-and-befriend instinct). Women offered themselves, to spare their children and their own lives. Once the Russians had moved on again and a large number of German soldiers had returned home traumatised, this created great tensions. Many raped women once again suffered greatly. The scale on which rape has been taking place in modern types of warfare has no parallel with prehistory. Because soldiers tend to fight far away from home, they can often get away with it, and at times are even encouraged by their officers, as during the break-up of Yugoslavia or amongst the foreign jihadists who are fighting in the IS-controlled areas of Iraq and Syria. In ancestral times, the attacked tribes would immediately retaliate. Now, the perpetrators are anonymous, the victims just as defenceless and any children stigmatised.

War over property

The agricultural revolution and subsequent population explosion had a significant impact on the scale and complexity of wars. Agriculture gave rise to something that had not been there two million years previously: property. Property had to be defended against people who would like to appropriate it. And so a soldier class appeared who came to protect belongings. Villages were given extra protection to make defending them easier. In every village or early city, in Turkey for instance, we see walls around houses. One of the first settlements in the world, Çatalhöyük in Turkey (7400 BC), has defences. Later, fortifications and castles were built. The size of communities exploded, and with it social complexity. Professions, trades and specialisations sprung up. Someone became a farmer, another a clerk, a third person an artisan and the fourth a soldier. Waging war became a profession.

Walled villages became states, with their own administration and economy. Old social orders disappeared and the group people belonged to became bigger and bigger and ever more abstract. Whereas in ancestral times the interest of individuals towards the group produced coherence, in an ever-expanding society the role of religion and patriotism grew in importance to create cohesion. War on account of abstract issues such as religion or nationality was an unknown phenomenon in prehistory, but became increasingly *de rigeur* in later times (the crusades fought by the European knights in the Middle East are a good example).

Warfare also increased in scale and became technically more refined, leading to all kinds of mismatches. The traditional three Ss, primordial motives for our ancestral conflicts, were greatly magnified by the population explosion. More people meant more sex, more status and a bigger salary for the winners of wars. The stakes were raised, which meant more interests had to be served. In terms of mismatch, the spoils of war became an exaggerated cue to which young men were attracted.

If one walled village occupied another, the booty (the earnings) were many times greater than during prehistory: large plots of land were taken, livestock was confiscated and many potential brides kidnapped. The status of war victor only rose. Winners became warlords, later barons, and even kings (the present Dutch king is a descendant of a war hero, William of Orange) and emperors who were in a position to maintain an even bigger army. The more soldiers someone was able to deploy in battle, the greater the chance of victory; the harder you fought, the greater the chance of large spoils.

Warfare became an increasingly effective means to acquire property. The demand for soldiers became so huge that many men fought alongside the highest bidder for decent pay. These mercenaries (a mismatch, because they did not exist in ancestral times when people always fought to protect their own group) were numerous on account of the population explosion. A well-known phenomenon during the Middle Ages was that of the first son pursuing academic study, the second becoming a skilled manual labourer and the third a soldier. Mercenaries were not in the slightest bit interested in for whom or for which objective they fought. A clear link between money and violence began to emerge: the richest rulers were able to afford the greatest number of soldiers and thus achieve the greatest number of victories (a principle that continues to live on in global football, a modern manifestation of ancient tribal disputes and the male warrior effect).

Peace for our time

Despite the growing military might of villages, cities and states, aggression gradually diminished in relative terms. Whereas in pre-agricultural times, around 25 to 30 per cent of the male population died as a result of violence, Harvard psychologist Steven Pinker believes that this subsided after the advent of agriculture. Now the figure in Western Europe and the US is less than 1 per cent. New, much more complex societies meant ever stricter standards and rules for the use of violence. States de-armed citizens and no more unnecessary risks were taken to attack neighbouring communities. Religions that advocated peace rose up, at any rate within people's own local communities. The bellicose instincts of young men especially were channelled and regulated much more in agricultural societies, and these instincts have emphatically come to the fore in modern society in the shape of hooligans, (motorbike) gangs, criminal syndicates, the mafia, monopolists, and ... CEOs.

Compare the pacification of humans against pigeons, who peck each other frequently with their beaks. Aggression is remarkably widespread amongst pigeons. They peck and peck each other to fight over hierarchy in the group. That's why biologists refer to this phenomenon as the pecking order, not to be confused with the length of the male member. Scientists once gave pigeons heavier weapons, by attaching knives to their legs and needles to their beaks. Aggression between the test pigeons dropped almost instantly, because the risk associated with violence was too great. The Cold War was based on this principle: if both the US and the Soviet Union had weapons of mass destruction, they would not attack each other. This geopolitical strategy was also known as MAD (Mutually Assured Destruction).

But there is another side to the story. If we do not look at the relative number of fatalities from war, as Pinker does, but at the absolute numbers, then we see an enormous rise in the number of casualties worldwide. This obviously has everything to do with the population explosion and the increase in lethal weapon technology, creating a mismatch. While the likelihood of dying in a war is many times smaller than before, in absolute terms, many more lives are lost than in earlier times. For the family and friends of the war victims statistics do not count, only individuals. This also applies to the hawkish leaders who send them to their deaths. The Russian dictator Stalin spoke these legendary words: 'The death of one man is a tragedy, the death of millions a statistic'.

The motherland is missing you

With the passing of time, the way in which war was waged changed. In ancestral times people fought with their fists, later with sticks, stones and axes. After the agricultural revolution, technology became an increasingly important part of warfare. Societies militarised, resulting in a large amount of effort being devoted to the manufacture of weapons. This, in turn, gave rise to an arms race. Weapons became ever more inventive and numerous, and the emphasis on crude, physical strength shrank. Moreover, people had to fight in wars further and further away from home for increasingly abstract ideals. This led to several mismatches.

The concept of 'love of one's country' (patriotism) has been invented purely for warfare and hints at mismatch. No normal biological organism would want to run great personal risk in order to protect an abstract concept like country or a state (only some bellicose ant species manifest something resembling this, but they do this while protecting hearth and home – at least their own nest). Humans have an instinct to protect immediate family and community in danger, but not to defend a religious objective or an abstract nation state whose inhabitants we do not know personally.

How have cultural innovations been able to counterbalance this fundamental mismatch between fighting for your family or your country? First of all, modern humans have a 'symbolic mind' at their disposal, the ability to make connections with genetic strangers, people who are not family or friends, but with whom we do have something in common. This symbolic brain is extremely useful, because it enables humans to join larger social networks. We have already seen how convenient this is in times of war, or when there was a threat of drought or water shortages. The larger your social network, the greater your chance of survival (this is probably what did it for the Neanderthals. Their networks were too small to survive in times of food shortages). The symbolic brain enables us to trust other people implicitly on the basis of a symbolic shared feature – such as belonging to the same church, nationality or football club. Patriotism is a corollary of this.

Rulers, politicians and generals can manipulate our symbolic brain by instilling in us the idea we belong to a group that needs defending. Millions of people have been united by the abstract notion of a homeland, even though they have no genetic interest whatsoever in fighting or dying for each other. Patriotism and religion are cultural constructs, which are intended, amongst other things, to organise and defend society. How many soldiers have died for something that had nothing to do with themselves or their genetic relations? A good example is both the World Wars of the last century, which saw millions of casualties.

But do people put up such a tough fight for abstract things like religion or one's country? Studies show that, in war situations, soldiers still primarily fight for their band of brothers, their small group of fellow soldiers. Hatred of the country and the politicians who send them off to war is huge. It also turns out that, despite the growing numbers of people taking part in combat, quite a number of soldiers do not play an active role. Research shows that only 20 per cent actually use their gun for shooting in battle. No wonder Pinker perceives a percentual decline in violence. Our instinct enables us to risk our life for our family or community, but perhaps to make less of an effort for our people and country.

In *Bowling for Columbine*, filmmaker Michael Moore shows that it is especially the disadvantaged in American society who fight the hardest and are most likely to die in battle. For a large group of disadvantaged young men, the army is one of the few opportunities to make something of life, acquire heroism and status and, when returning from battle, build a social existence and have children that will lead a respectable life. We see something similar in traditional nations. Bellicose Plains Native Americans make a distinction between peace leaders and war leaders, and these roles are fulfilled by different people. Peace leaders tend to be older and come from well-off families within the tribe, while the war leaders are younger, wilder and come from less-endowed families. So it appears that war continues to give disadvantaged young men a means to acquire status.

But the likelihood of becoming an actual hero and making a difference has decreased somewhat in modern complex wars between countries. In ancestral times, you would be instantly recognised as a hero; the five of you set out and those who returned, at the very least, had done nothing disastrous in the battle. Modern warfare, partly as a result of military specialisation, involves so many people that it is almost impossible to stand out above the rest. What's more, weapons are much more lethal these days, so walking out in front in order to become a hero tends to be a bad plan. In sharp contrast to the 4486 American soldiers who died in Iraq stand the tiny number of Medal of Honour recipients.

War at a remove

An important facet in the cultural development of warfare in recent history is that the distance between the fighting groups – and the perpetrators and victims – has been growing ever larger. We saw this in the drone example at the beginning of this chapter. This increasing distance produces mismatch with far-reaching consequences. The entire weapon technology is aimed at widening the distance between perpetrator and victim, in order to reduce the risks to the aggressor. Moreover, by creating distance, empathy for the other party dwindles which in turn leads to an increase in the likelihood of atrocities and sadism. Greater distance means less empathy, an absent cue during modern conflicts with all its attendant consequences.

The psychologist Daniel Batson illustrated this in a study in which he told students in his lab that there was a student in another room who had to answer various questions, and every time he made a mistake would be given an electric shock. The researcher asked each student whether he wanted to swap with the victim. The reply depended on the similarity between the two people. If the victim was described as someone who was the student's equal in some respects – for example, someone with the same taste in music, or political views – then the student did not mind swapping places. In other cases he did mind.

When our ancestors were still travelling across the savannah, it was, as we saw, important to mobilise as many men as possible to – extremely cowardly – bump off individuals from a hostile tribe. Physically, men should get the better of one or two (unsuspecting) intruders. But when the attackers began to use spears and axes to defend themselves, it became important to stand further apart, out of self-protection. Swords emerged, and later guns and cannons, weapons that became more and more dangerous, used from an increasingly greater distance. And with that large distance, the ability to feel another person's pain decreased.

The natural connection between perpetrators and victims which had been the deciding factor for hundreds of thousands of years, got lost completely. Thanks to the arms race, it was possible to catapult enormous boulders at castle walls and kill enemies at a huge distance, as well as to bomb entire cities. The number of warriors an army had no longer counted; what mattered was destructive power and technological edge. There is an interesting anecdote about how technology can affect old-style warfare, chronicled by Jared Diamond. He describes how a tribal chief in New Guinea saw an airplane land for the first time, filled with missionaries. The chief asked whether he could fly in the bird and the missionaries agreed. Before getting on board, the man wanted to load up a pile of large rocks. The missionaries were dumbfounded, until he explained he needed the rocks to drop from a great height on his enemies in the neighbouring village.

With increasing frequency, wars began to be waged far away from home, whether it was the Romans who hit a wall of hostility in Scotland, or Napoleon's soldiers who froze to death in the Russian arctic fury. This mismatch often ended up badly. An important biological mechanism is that of the home advantage. A robin protecting his garden puts up a greater fight than a robin invading a territory. The aggressor can try elsewhere, but a defender fights for his nest and offspring. Which is why many wars are won by the country with the home advantage. Waging war in nowheresville tends to end in tears. Whether Americans in Vietnam, Russians in Afghanistan or Germans in Russia, the further soldiers are away from home, the less motivated they seem to be to fight.

The role of leader as mismatch

Waging war on a large scale is most effective when a group is organised in a strictly hierarchical fashion, with delegating leaders at a distance. In ancestral history, leaders went first and led from the front, but the generals who sent their soldiers into the trenches and battles during World Wars I and II did so at a safe distance from the clashes. From an evolutionary perspective, leading from the back was a new phenomenon because, by tradition, the head of the battle was a better place to make assessments than miles away. Leading from behind the lines also caused millions of unnecessary deaths and historical blunders. We know from World War I that, from behind the front, British generals often took the erroneous decision to send their soldiers out of the trenches to attack the Germans, resulting in hundreds of thousands of needless casualties. The fact that the leaders did not experience the consequences of their mistakes for themselves is a mismatch. If there is a chance you will be killed, you think twice before you launch an attack.

In modern warfare, it is the leaders who run the least risk instead of the most, as in ancestral times. We are astounded when a general is prepared to take a personal risk or expose his descendants to risk. People were full of praise when Dutch General Peter Van Uhm lost his son in Afghanistan. This was an intensely personal drama, with the unintended consequence that Van Uhm revealed himself as a leader who was prepared to bear great sacrifices, just like past leaders. This has given him an unprecedented high status, and he is the best-known Dutch serviceman – if he had aspirations to enter Dutch politics, he would stand an excellent chance.

Psychological war complaints as mismatch

In 1980, the third edition of the *DSM*, the *Diagnostic Statistical Manual of Mental Disorders*, described the symptoms of post-traumatic stress disorder (PTSD). This is a psychological disorder which is classified as an anxiety disorder. The recognition of PTSD was largely associated with the Vietnam War, from which many veterans had returned unhinged. A remarkable number of young men became addicted to drugs, displayed maladjusted and often criminal behaviour, had hallucinations and frequently withdrew – armed – from society. Once the disorder had been included in the *DSM*, medical science threw itself at the condition. The number of publications in scientific journals exploded, from forty-three in 1980 to seventeen hundred in 1999. Since then, more than six thousand articles about the topic have appeared, with PTSD no longer solely relating to the problems caused by war, but to all types of trauma from aggression.

That war was an important cause of stress had been long known in world literature. Literary historians suggest symptoms of this psychological disorder can be found in the *Epic of Gilgamesh*, Homer's *Iliad* and the writings of Herodotus. The latter's *Histories*, for instance, feature a brave warrior who has to cope with blindness following an intense fight. Many examples of war neurosis exist in old books. In some Native American tribes, warriors have to live outside the group for a while when they have killed enemies, before being included into the community again. The objective is to ensure that they have recovered mentally in order to take part again in ordinary daily group-life peacefully.

PTSD can arise when people find themselves in life-threatening situations or have to contend with severe physical injury. There are many symptoms, including tiredness, shortness of breath, palpitations, sweating, chest pain and even fainting. PTSD is often accompanied by recurrent depression, anxiety, irritability, re-experiencing of the stressor, difficulty in concentrating, problems falling asleep, aggression issues, substance addiction, problems in forming and maintaining long-term relationships and social isolation. Here, too, our evolutionary history continues to affect us. Even if the drone pilot who creates long-distance casualties in the Nevada desert does not run a personal risk, it is still psychologically stressful to kill people, and probably even more stressful when you have not faced eyeball to eyeball the person who was actually out to finish you off. At one point, waging war was a very personal thing. You could see, smell and feel your enemy: now they are green specks on a screen.

War heroes continue to have a certain status, but for a long time veterans have not received the praise and financial compensation they are due. The story with which this chapter opened is based on that of Brandon Bryant, a US Air Force operator who, in 2012, lifted the lid on the time he helped to kill people from a distance. During the six years that he worked as a drone pilot, his squadron was responsible for the deaths of no less than 1626 people.

When Bryant had killed a child in one of the attacks, he wrote in his diary: 'On the battlefield there are no sides, just bloodshed. Total war. Every horror witnessed. I wish my eyes would rot.' Partly because of this incident he began to cut himself off from his friends. He suffered from chronic bad temper and hardly slept. One day he collapsed: he fell over and vomited blood. Doctors diagnosed he suffered from PTSD. He is no longer in the army, but living in poverty. No war hero, in other words.

The atom bomb and other weapons of destruction

The ultimate consequence of the arms race – and the most alarming mismatch – is the development of the atom bomb, a weapon that has in fact only been deployed twice and has been such a terrifying deterrent that no one has ever dared to use it since, aware that dropping a nuclear bomb equals suicide. Military experts and political scientists talk of a MAD-strategy, which we discussed earlier: guaranteed mutual destruction. These days rockets can see to it that both the attacker and the defender are eliminated and this idea has – for the time being – proven sufficient to keep the peace.

But it is still not a comfortable situation, least of all with a current Russian president with an ego problem.

Army bases around the world house enough nuclear weapons to destroy all life on Earth (and several times over if that were possible). At present, there are seven states with nuclear weapons. Two countries (Israel and North Korea) are believed to possess nuclear weapons, but the truth is unclear. Seventeen further countries, including the UK, France, and the Netherlands, have nuclear research programmes which could lead to the development of a nuclear weapon. In a word, if the aforesaid power-mad Russian president, the authoritarian Chinese party leader or the North Korean dictator gets the idea in his head, it could be the end of the blind process of organic re-arrangement of elementary components. It hardly needs stating; but our primitive brain finds it difficult to grasp that there are global powers in possession of weapons capable of wiping out a large part of humanity in one fell swoop. If our brain took this threat truly seriously, we would have ensured that countries like Russia, North Korea and Pakistan were nuclear-free.

Sport as a substitute for war

Where our evolutionary warring past manifestly lives on is in our fascination for sporting competitions. We could see team sports as a ritualised form of warfare, a positive mismatch, as the conflict rarely claims fatalities. The rules all players in a game have to adhere to ensure that the battle acts do not deteriorate into a true war. Just very occasionally it comes to grief.

On 10 October 2014, a European Football Championship qualifying match between Serbia and Albania was underway in the Serbian capital of Belgrade. Out of nowhere, a drone appeared over the pitch with an Albanian flag attached to it. The Serbian player Stefan Mitrović tried to grab the flag, whereupon a fight broke out as some of the Albanian players began to defend their flag. At this point the referee suspended the match. After the game, the brother of the Albanian prime minister was arrested, on suspicion of operating the drone from the stands.

The root of these skirmishes was the declaration of independence by Kosovo, which had been declared an independent republic by the Albanian majority in 2008 – against the will of Serbia.

When missionaries introduced football in Papua New Guinea this initially led to many scuffles as well, sometimes ending in full-blown tribal war. Thankfully a catholic solution was found to this problem by introducing a new rule to the game. All matches had to end in a draw, which meant there were times the teams had to keep on playing for a long time.

Sport is the channelling of our primitive instincts for warfare: we wear tribal colours, we fight for trophies, we experience home advantage and during play many hormones are released that are comparable to the hormones released when fighting. These hormones pump not only through the players on the pitch, but also through the spectators (as we will see in Chapter 9). Studies show that, amongst football supporters, changes in cortisol (the stress hormone) and testosterone (the status hormone) can be measured. In fans of both teams the cortisol level rises during the game. In fans of the losing team the testosterone level plummets, while that of the winning team soars.

In the times of our ancestors, wars did not have many spectators, so there is a case of mismatch as a result of exposure to a fake cue. In a conflict between two tribes, everyone took part in combat action, and those who did not were socially excluded or worse. Nowadays we sit in stands to watch clashes, often accompanied by stress, raised blood pressure and palpitations. Taking part in a sport may be healthy, but watching sport carries serious health risks. Our fight-or-flight response – useful when we are engaged in a pitched battle – is also activated in a full football stand where we have nowhere to go when our team threatens to lose. It is not for nothing that football violence spreads through the stands with some regularity. The most famous football conflict is possibly the so-called 'football war' between El Salvador and Honduras. After riots between supporters broke out during a game between the two sides in 1969, the countries fought each other for four days. Two thousand men lost their lives in this military conflict.

Break a lance for peace

It should be clear by now that modern warfare has created all kinds of mismatches.

At the end of this chapter we would like to turn our minds to the question of how we might be able to use the mismatch theory to promote peace. What should we do with all these male warriors? We could start by doing nothing. We could simply accept that there have always been generations of young men who find it exciting to take part in organised violence, whether they are hooligans, street gangs, the mafia, or IS. It is biology. We could stand by and watch how at some points in time a society turns into a version of *Mad Max*. And if we do not want to have them in our society, we could arrange a war for them and make sure they go and fight for us a long way from home.

The question is whether they would want to do that. As we have seen, soldiers primarily fight for their band of brothers, many hate the country or politicians who dispatch them, and a large proportion does not even take part in active combat.

Doing nothing also means that we, with our Stone Age brain, have to learn how to handle weapons which are becoming ever more lethal, with the distance between perpetrator and victim growing ever larger. In order not to run any risks itself, the American army is in the process of deploying robot soldiers to attack the enemy. But a widening distance leads to a decline in empathy, and consequently an increased likelihood of excessive violence and erroneous decisions. One of the reasons whistle-blower Bradley Manning (now Chelsea Manning) decided to leak to WikiLeaks secret video recordings of an extremely violent American helicopter attack on Baghdad was the excessiveness and inhumanity of the action. This act landed Manning a thirty-five-year jail sentence (although she was since pardoned). Another consequence of this brute force is that the enemy also starts to fight more aggressively in order to restore the imbalance of power. So, in asymmetric wars, violence will only increase, for example through combat with suicide platoons or by carrying out bioterrorist attacks.

A second solution to mismatch could be to make peace more important. Our ancestors had adaptations to wage war with other groups, but they also had adaptations to make peace. The Yanomami are not continuously at war either. Sometimes there is a conflict, but the villagers tend to live in relative peace with each other.

How can peace be safeguarded and what lessons can we learn from our evolution? One of the things that brings communities closer together is the exchange of marriage partners, which results in neighbours' genetic interests mixing. This also happens amongst hunter-gatherers, partly to enlarge social networks. During the Middle Ages, this was *the* way to unite royal families from different realms. One of the most successful ways for societies to counter prejudice and xenophobia is, not for nothing, mixed marriages with children.

One option is to have women play a more prominent role in politics and the military. As previously shown, there are generally few evolutionary advantages attached to warfare and often only disadvantages (rape, for example). As leaders, women will therefore be less quick to launch a war. Studies indeed show that women are slower to initiate a war when, as leaders of a country, they have to solve a simulated crisis in a lab situation. Margaret Thatcher and Golda Meir are female leaders who started wars, but they appear to be the exceptions to the rule. It turns out that in lab-simulated wars, female participants opt for attack far less quickly than men and continue to negotiate with the 'enemy' for far longer before going to war.

Why don't we send so-called 'UN women's battalions' to the world's hotbeds? With weapons in their hands women are as deadly as men, but they have greater empathetic powers and arouse less aggression. The same could apply to men with feminine traits. The problem is that in a conflict with another country, we prefer masculine men to do the fighting for us.

Neuroscientists who work for the US Military are also investigating whether it might be possible to spray substantial doses of oxytocin, the bonding hormone, from aircraft over war zones. Enemies would become friends and break off action.

Likewise, women could instigate a collective sex strike as a protest in order to change the minds of men who are waging war. This possibility was presented in a comedy in Greek Antiquity, *Lysistrata* by Aristophanes, as a solution to the protracted battle between Athens and Sparta. But sex strikes have indeed happened. There had been fighting between the separatists and the Philippine government army on the Philippines' second largest island, Mindanao, since the 1970s. The United Nations High Commissioner for Refugees (UNHCR) provided the battling village with emergency aid, but this had little effect. Until the warriors' wives threatened to withhold sex. Men who went off to fight would no longer receive a loving welcome in their marital bed. This turned out to be a more effective remedy than the UNHCR food parcels and fishing-nets.

Similarly in Africa, Togo women once called a sex strike. Women belonging to the Togolese human rights organisation Red Togo had copied the action from Liberian women, who in 2003 would only make love again if their men made peace with their enemies. 'We have many means to force men to understand what women want in Togo,' a lawyer said at a women's rally in the capital of Lomé.

Another way to solve mismatch in warfare would be to make leaders discover more for themselves what it is like to fight. Leading from the back has resulted in leaders no longer experiencing personal risks. They barely know what they put their soldiers through. Perhaps it is an option to force generals to take part in foot patrols and to lead in battle, in other words to lead from the front. That will teach them!

Finally, peaceful substitution for warfare could be pushed further. We know from studies that in terms of hormones (a combination of testosterone, cortisol and oxytocin) and group bonding, doing battle with a band of brothers has a comparable effect to a bout of paintballing or engaging in an exciting sports match. The battle being fought is a symbolic one without casualties. More sport means that a greater number of people can participate simultaneously. In order to prevent cheating and match fixing, it might be advisable to work with female referees only. Another substitute for warfare is already happening at home, where boys play *World of Warcraft* on their PlayStations or online on their computers with team members from different parts of the world. Doing something with people online across the world expands people's empathy and mutual understanding and reduces the chances of war. So, China and Russia, if you also read this: make the internet freely available.

CHAPTER 8

Waiting for the ice to melt

Café Victor serves the kind of meal that is being offered in better restaurants the world over. Various exquisite dishes make their way from the kitchen: beautiful starters and a pad thai with prawns from Norway, noodles from Italy, lemongrass from a Dutch greenhouse, peanuts from the American state of Georgia and fresh red peppers from Mexico. The whole world laid out on a plate.

The dining room is warm and cosy. The white French house wine is chilled and the red from South America a pleasant room temperature. Geography teacher Peter Rodgers from Leeds and his family are savouring the lavish dinner. If we did not know any better, we might think this fusion restaurant was somewhere in New York, Buenos Aires or Copenhagen. What sets Café Victor apart is its view, perhaps the most spectacular view on the planet.

Peter Rodgers and his family arrived that afternoon after a three-hour boat trip from Ilulissat in Greenland. Since then, they have been gazing in wonder at what Café Victor overlooks: the Eqi Glacier, a glacier wall stretching five kilometres and rising up a hundred metres at the front and two hundred and fifty further back.

The restaurant is named after the French explorer Paul-Émile Victor (1907–1995). It was partly his research that led to the discovery of three islands beneath the ice cap which later was to house the restaurant named after him. Peter had been there before, twenty years ago, on a student-expedition with his university.

Then, the researchers had put up tents alongside what was known as 'the Frenchman's cabin', the remainders of the cabin Paul-Émile Victor used during his expeditions to the Eqi Glacier and the area called Inland Ice Sheet. This ice sheet is made up of snow, snow and more snow, in such thick layers that the snow has turned into ice. As a result of the force with which the snow has been pressed together, flakes and air bubbles have formed inside. When chunks break away from the glacier – a regular occurrence – this makes such a tremendous noise that travellers have compared the crashing to the end of days.

In 2001 Glacier Lodge Eqi, fifteen huts in a semi-circle around Café Victor, was built. The cabins, some with hot water and underfloor heating, look out onto the glacier. Visitors can admire the collapsing ice from their rooms, the terrace or the restaurant and allow themselves to be overwhelmed by the thunderous concert. Or as the folder puts it: 'A very meditative way to spend time and enjoy the fireworks of nature.'

The Eqi glacier wall shifts around nine metres a day. Peter's family see immense chunks of ice ('taller than Big Ben', Peter impresses upon his children) breaking off from the face of this ice horizon. The glacier is in constant motion. One of Peter's children says it is like looking at a film in slow motion: when a gigantic chunk peels away from the glacier it initially sinks into the ocean, excruciatingly slowly at first, then forming a tidal wave reaching metres high over the water, thick with shards of ice and icebergs.

As Peter is taking a moment to smoke a cigarette on the restaurant's terrace, he and his family are taking in the imposing mass of ice. Peter, who became a teacher for good reason, explains that the Greenland ice cap has a volume of water that could make the entire ocean rise by seven metres if the ice were to melt.

'And is the ice melting?' his daughter asked, concerned.

He has to confess it is, but does not add that he is extremely alarmed at the state of Eqi and the rest of the ice in Greenland. Since the time he was part of a team measuring the ice and melting conditions, a great deal has changed. Greenland, which is situated in a relatively warm climate, has a vast ice cap which, at its edge, has an average temperature of around zero degrees Celsius. This temperature makes researching changes in the climate easier, because ice cannot get any warmer than zero degrees. A minor change in temperature will reduce the ice cap immediately, Peter tells his children. At the time, researchers believed that the Greenland ice cap was more or less in equilibrium, or that as much new ice was generated as was melting away from the edges.

Peter is now witnessing with his own eyes how the ice mass has indeed shrunk, and by some way. He had already read various reports and investigations about this (99.9 per cent of climate change research indicates that climate change is in fact happening), but seeing the changes for himself it really hits him (and perhaps the red wine plays a part as well). The Eqi ice wall which had bowled him over when he was a student has not lost its immense power, even if its expanse and height have visibly shrunk.

'Is that bad?' his son asks.

'Not just yet,' Peter replies. 'It is estimated that sea levels will rise globally by about thirty centimetres a century, so we can cope with that for now. But the faster the ice melts, the faster sea-levels will rise as well.'

'So in about a hundred years' time this ice may have melted for good?' his daughter asks. 'Then the view won't be nearly as good . . . '

The three of them gaze across the water and cannot imagine that something as impressive as that could ever disappear. Then Peter's wife knocks on the window of the restaurant. Café Victor's Danish chef has served the sweet course. Peter and his children rush inside the warm room, where plates display delicious desserts: an extra-vagant portion of hot chocolate sauce from California is melting a path across the home-crafted vanilla ice cream made with cream flown over from the Swiss Alps and sugar from Brazil. As they take their first spoonful, their blissful groans momentarily drown out the glacier's roar.

What did the world look like in earlier times?

In the first chapter we discussed EEA, the environment of evolutionary adaptedness. Our genes were formed on the African savannah, which was not exactly over-populated with fellow human beings. It is estimated that up to a few thousand years ago, no more than one million people lived on the entire planet (now there are all but 7.5 billion). Every ancestral human had some 150 square kilometres at his disposal. By way of comparison: the population of the Netherlands currently stands at just under five hundred people per square kilometre. In other words, people hardly got in each other's way.

Our ancestors were nomads; they moved from place to place in pursuit of food and drink. They lived purely in the here and now. It was functional to think in the short term because you did not know if you would still be there the following month, let alone the following year. As we explained earlier, our ancestors lived in a verdant landscape, rather like parts of the UK or France.

The question is whether, at the time, humans were responsible for sporadic environmental pollution (the topic of this chapter). Locally, this will certainly have been the case. Everyone knew where to relieve themselves: not in the same place where people ate or slept. It was also seen as good sense to clear away the carcasses of consumed animals before predators came for them.

Several times a year (some scientists think as often as eight), our ancestors moved, to a different camp, to a different feeding place.

They may have developed a tentative form of 'sustainability initiatives'. They must have had knowledge of animals' growing cycle; which animals to leave alone – the mothers and their young – and which animals would be a good catch. Once every so often, areas of woodland would be burnt down. Only relatively few trees were felled; wild crops were harvested, but not cultivated on a large scale. There were so few people in such an expansive area that their behaviour and way of life had little impact on the flora and fauna. In short, humans played no significant part in the life of the planet.

The myth of the ecological savage

One of the myths of the present time, partly propagated by philosophers and environmental activists, is that our ancestors were extremely 'environmentally aware'. Famous statements suggesting that 'man was the keeper of nature and everything that lived' have been made by various people including the Native American Chief Seattle (a guru from north-western America).

One of our parental homes had a copy of parts of the speech Chief Seattle is said to have given in 1854 in reply to a territorial governor, parts that Al Gore would later include in his book *Earth in Balance* (1992):

> How can you buy or sell the sky? The land? The idea is strange to us. If we do not own the freshness of the air and the sparkle of the water, how can you buy them? Every part of this earth is sacred to my people. Every shining pine needle, every sandy shore, every mist in the dark woods, every meadow, every humming insect. All are holy in the memory and experience of my people . . .
>
> If we sell you our land, remember that the air is precious to us, that the air shares its spirit with all the life it supports. The wind that gave our grandfather his first breath also received his last sigh. The wind also gives our children the spirit of life. So if we sell you

our land, you must keep it apart and sacred, a place where man can go to taste the wind that is sweetened by the meadow flowers.

Will you teach your children what we have taught our children? That the earth is our mother? What befalls the earth befalls the sons of the earth. This is what we know: the earth does not belong to man, man belongs to the earth. All things are connected like the blood that unites us all. Man did not weave the web of life, he is merely a strand in it. Whatever he does to the web, he does to himself.

With this speech, Chief Seattle took a stand against the white intruders who had come to America to kill off bison, buffaloes and Bodéwadmi Native Americans. 'I have seen a thousand rotting buffaloes on the prairie left by the white man who shot them from a passing train,' he said in his speech. Unfortunately, all his words proved to have been invented. The railway tracks to which the Native American chief was said to have referred, did not yet exist in his time. His fine words, recorded by someone who did not speak his language, were comprehensively rewritten and adapted by a Hollywood scriptwriter during the last century to convey a political message. The myth of the sustainable hunter-gatherer, the Native American, meshed neatly with the growing environmentalism of the 1960s and '70s.

Many activists and politicians went back to the work of Enlightenment philosophers, who contended that ancestral humans lived in harmony with nature. The basis of this philosophy was laid during the eighteenth century by the French philosopher Jean-Jacques Rousseau, who thought that our primitive ancestors had a more positive outlook and fared better in life than their over-evolved eighteenth-century counterparts. In terms of their refinement, primitive people occupied a middle position between wild animals and decadent modern humans. As he wrote in 1754:

Hence although men had become less forebearing, and although natural pity had already undergone some alteration, this period of the development of human faculties, maintaining a middle position between the indolence of our primitive state and the petulant activity of our egocentrism, must have been the happiest and most durable epoch.

The more one reflects on it, the more one finds that this state was the least subject to upheavals and the best for man, and that he must have left it only by virtue of some fatal chance happening that, for the common good, ought never to have happened. The example of savages, almost all of whom have been found in this state, seems to confirm that the human race had been made to remain in it always; that this state is the veritable youth of the world; and that all the subsequent progress has been in appearance so many steps toward the perfection of the individual, and in fact toward the decay of the species.

The decay of the species, that was the point for Rousseau's contemporaries. Because of these words and ideas, Rousseau is often associated with the term 'noble savage' (which he did not use himself), an ancestral human who treats his living environment with respect, in contrast to his contemporary counterpart who has been completely corrupted and abuses nature for personal gain in a way that borders on the disgraceful.

Alas the idea of the ecological noble savage is a persistent misapprehension, because early humans were not in the slightest bit sustainable, and concepts such as 'living in harmony' meant nothing to them. Extravagance simply did not exist, as there was nothing to accumulate in the first place. Nor did people have technologies to deplete nature; there were no guns to shoot animals, no huge fishing nets to plunder the oceans. If they had had these, their large brains would have most certainly have prompted them to use them.

247

Anthropological studies show that, in traditional societies, there is no relationship between how sacred nature is to them – attested by the number of rituals – and how sustainable their behaviours is. Sustainability is solely determined by factors such as population density and the presence of modern technology.

To recap: our ancestors lived in a non-cultivated natural environment in which they roamed from place to place, and gave no thought to what they left behind. They were impulsive, lived from day to day and did not concern themselves with 'the common good'; nor did they regard 'the environment' or 'climate change' a problem, because it was not a problem. Our ancestral brain gets in the way of us solving the global sustainability issues facing modern humans. The disaster scenario sketched by environmental scientists Joel Heinen and Bobbi Low from the University of Michigan based on knowledge about human evolution is: 'Natural selection has shaped all living organisms to exploit resources effectively. Our human problem is that, through our cleverness, we have created a novel evolutionary circumstance. We now have such technology that the very behaviours which we evolved to do well, are those most likely to ruin us.'

Our not-so-green brain

The main topic of this book is that our primitive brains are not well-adapted to modern life, resulting in a mismatch. Five aspects of our ancestral brain conspire together to create a horror scenario for the ecological damage we as a species are inflicting. Formerly, these five aspects enabled us to survive and flourish in the savannah, but in our modern environment they produce mismatches. This has resulted in massive environmental problems heading towards us like a boomerang from hell – problems that we have no idea how to solve.

First of all, our primitive brain's main focus is our own welfare, or at any rate that of our families. We are egocentric and selfish and we put up with the negative feelings these notions arouse in us. This is the starting point of the selfish gene theory advanced by Richard Dawkins. Our own well-being and that of our close relatives is more important than that of genetic strangers. How much money are we willing to spend on someone we do not know? A great deal of psychological research has been conducted into how people behave when asked to distribute a particular sum of money. One way of measuring generosity is the so-called 'dictator game', in which participants are given money to divide between themselves and other unknown participants (there are many variants of the game). On average, people give away only 28 per cent of the money they receive. Women and the elderly give away more, and most money is given to people who have ended up in financial difficulties through no fault of their own.

But the majority keep most of the money for themselves. They give away only a small proportion of it to a stranger, even if the latter needs it more. Studies also show that people become more generous when the recipients are family members and friends instead of strangers. We are prepared to give more to colleagues than to strangers, more to friends than to colleagues and more to relatives than to friends. In ancestral times, we were not interested in tribes on the other side of the savannah, simply because we had no dealings with them. The far side of our camp was where our world ended, so why should we care about the world outside our camp?

The consequence of this primeval drive towards self-interest is demonstrated by the tale of 'tragedy of the commons', a concept in game theory. The mathematician William Forster Lloyd described the problem as far back as in 1833, but it became widely known in the 1960s due to the efforts of ecologist Garrett Hardin. 'Commons' are common land on which herders can collectively graze their cows or sheep. Each common has a set size and only so much grass. If a farmer adds an extra animal to his stock, he benefits (an additional cow or sheep means more milk or wool), but the other farmers are disadvantaged (because there is less grass left for their animals). They in turn will try to increase their stock with new animals, resulting in overgrazing and depletion of the common, which means all farmers lose out. One individual farmer's benefit leads to all the farmers' tragic ruin. We see this tragedy of the commons today in oceans that have been fished dry and in polluted living environments. We have too many children (genetic self-interest) and as a result the global population is growing, something the planet cannot cope with. It is in everyone's genetic interests to have more children, but the environmental costs have to be shouldered by everyone in society, including people who are childless.

Secondly, our brain continues to focus on the here and now, instead of on the future. We want instant gratification. Many scientific studies have demonstrated the existence of that childish instinct, including the famous marshmallow test devised by psychologist Walter Mischel in 1972. Children were offered a marshmallow which they would be allowed to eat immediately. They could opt to hold off eating it, in which case they would be given a second one fifteen minutes later. 'If you put this one aside, you'll be able to enjoy a second later with pride.' Many children responded impulsively and grabbed the marshmallow then and there. Yet there were plenty of others who were able to hold off. Follow-up studies showed that these children were more successful in later life. It is likely that people with a high degree of self-control are better at dealing with mismatches. Self-control has been proven to be a useful indicator of how an individual will perform at school, cope with stress and function socially; whether they will be able to keep their weight in check, not smoke, make sound financial decisions and invest in their pensions; and whether they will have successful loving relationships.

The third trait is our inherited status-awareness. Our brain is geared towards acquiring status, because in ancestral times it was associated with all kinds of evolutionary advantages. Questions that were forever at the foreground of people's minds were: who came home with the tastiest gnu carpaccio? Who was the best hunter? Who collected the juiciest fruit? Who told the best origin stories? Who was the most popular in the group? These individuals would acquire the highest status, and with it a healthy share of sex and scarce food resources. This innate predilection for acquiring status means that people always want more, and to be better off than those around them; and leads to huge extravagance and wastefulness in today's economy. To measure the importance of relative status, the American economist Robert Frank gave participants in his study a choice of two worlds. In World 1 they worked for an organisation where they earned 50,000 dollars a year, with the average annual salary being 60,000 dollars. In the other world they worked for an organisation where they earned 40,000 dollars a year, with the other workers on an average salary of 30,000. Which world would they prefer? Those who care more about absolute status would choose World 1, because they would earn more. Yet most people opted for World 2. The conclusion is that people are geared towards wanting to do better than those around them, even if that means wanting things they do not need, or doing things that go against the general interest.

Our primitive brain's fourth trait is copying the behaviour of those around us. The motto is: the majority is always right. In ancestral times that was obviously a good strategy. If, when out on the savannah, Johnny set out in a different direction from the rest of the group, it might be the end of him. Following the group provides protection. This is known as the herd instinct. If ten people on the street are looking up at the sky, casting a glance upwards yourself is not a bad idea. Who knows, a plane might be about to crash right on top of you.

But this copying behaviour also means that, if we see people around us dropping litter on the street or not cleaning up after their dogs, we are equally inclined to follow their example. A classic social psychological experiment shows how people conform to the group, even when they know that the group is wrong. The conformity experiments by the American psychologist Solomon Asch revealed that people tend to conform to the majority of the group. In one such experiment, test subjects had to look at lines of different lengths. They were asked to say which was the longest. What the test subjects did not know was that all the other participants were actually part of the experiment. If the majority of these answered that a certain line was longer, the actual test subjects would follow suit, even if they knew it was wrong. It appears peer pressure is more important to us than logic.

A final trace of our ancestral brain's past is its primary focus on direct sensory experiences. This impacts on how we perceive nature. A bad smell would mean a member of the tribe had just defecated or that some food was giving off a strong odour. If we hit upon an unpleasant smell, we would avoid the area. This is adaptive behaviour. But that means we ignore information we cannot directly see with our eyes, hear with our ears or smell with our noses. So we may read about melting ice caps, burnt-down rainforests and bone-dry water reservoirs in the paper, but it does not affect us; our back garden is looking a treat, flowers are blooming and clean drinking water is flowing from our taps. Our brain thinks: what is all the fuss about?

In a word, how our brain works impinges on our interaction with nature and the planet. Traits that enabled our ancestors to survive on the savannah with few people and ample space are now very much working against us.

Camping is fun

Obviously, the fact that our brain has evolved in a natural environment has produced positive offshoots as well. Take our preference for camping holidays. We like to 'get away from it all' to engage in age-old essentials: where do I pitch my tent, how do I get food, where do I find wood for the campfire and how do I form a bond with other people in the queue for the latrines? The desire for camping is a primordial need resembling the activity of hunter-gatherers. It goes without saying we do without luxury when we are camping and inadequate hygiene increases the risk of infections, but we are not put off by this (this is a mismatch, but we will come back to that later). When camping, we do the same things our ancestors were doing for hundreds of thousands of years: we search out an elevated, safe spot for our tent (which does not get flooded in the first rain shower), we stake out our plot, we try and have friendly encounters with our neighbours ('Could I borrow your mallet, please?'), we light a campfire or barbecue (with the men roasting the meat and the women preparing the salad) and we look after each other's children ('shared parental control'). Camping is not a holiday, but a primitive urge. The Dutch are experts at this like no others. According to a survey by Statistics Netherlands in 2014, more than 3.6 million Dutch people annually venture out with their tent or caravan for a holiday at home or abroad. By comparison, it is estimated that about 1.2 million people in the UK go camping on a regular basis.

Many studies confirm that experiencing nature is good for us. People who look at nature photos instead of photos of urban scenes feel calmer and experience less stress. Gazing at green images makes us happier. In airports like Schiphol, bird and animal sounds are played to calm people. Brain research shows that this is indeed soothing. Even a tiny patch of artificial grass has a calming effect, because our brain does not distinguish between real and artificial grass. The same goes for other animal species. Deer and rabbits are regularly spotted on football pitches with artificial turf. Having nibbled a mouthful of the grass, they walk away in puzzlement. Likewise blue surfaces create a calming effect (we associate this with blue skies in which we cannot detect a shower).

The famous hospital study by physiologist Roger Ulrich from 1981 revealed that patients recovered more quickly after surgery when their room had a view of trees than when their room looked out onto a blind wall. Evolutionary scientists call our instinctive preference for nature biophilia. Researchers showed photographs of different scenery (urban versus rural versus desert versus mountains versus rolling green fields) to people from different cultures and all of them turned out to have a strong preference for the latter. This is probably the landscape of the EEA – the savannah – and we distil from this the message: this is where we can find food, hide, be on the lookout for danger without being seen ourselves. The kind of scenery we experience as beautiful and aesthetically pleasing today is a deep-seated preference that has shaped and kept us alive for hundreds and thousands of years.

Changing nature

How did our planet change after the introduction of agriculture and later, after the industrial revolution? And how did our ancestral brains deal with that change? Better put: What was the human impact on the planet? Did sustainability suddenly become a topic? Humans, who had been nomads, began to settle in villages and towns. In a relatively short space of time, the Earth's population mushroomed. Before the advent of agriculture, small groups would migrate to places where there was food, without worrying about how they left their surroundings. Our brain was adapted to life in an environment where there was in principle enough food for all. After settling in fixed places, people were suddenly confronted with the question of how to meet the needs of large numbers.

In several places on Earth, early farmers parcelled out plots of land on which to grow crops. Woods and natural vegetation were turned into meadows. After a while, people tried to domesticate wild animal species (buffaloes, pigs, goats) and keep them near their crops, so that they did not have to chase after them. In a very short space of time, agriculture became more intensive, irrigation channels were dug, rivers channelled, entire areas were cleared and burnt. Human impact on the natural environment grew exponentially, as did the world's population.

Whereas the small groups of nomads had had little if any impact on nature and the environment, this could not be said of farmers and city dwellers. They intervened in the natural environment and adapted it to meet their evolutionary craving for status, self-interest, wastefulness and extravagance. As a result of this, the impact on nature has been massive, as has (particularly in recent centuries): pollution, over-use of materials, depletion of agricultural land, overfishing, mass extinction of species, a shortage of fresh water for agriculture and consumption, plastic waste in the oceans and an alarming decline in air quality in many regions. The many tiny contributions by many individuals have collectively resulted in an irreversible problem – a textbook example of 'the tragedy of the commons'.

For instance, we are selfish large-scale consumers of energy and water. Our short-term thinking leads to impulsive purchasing behaviour: we buy all kinds of things that we do not actually need to keep ourselves alive, and much of this ends up at the rubbish dump. Our craving for status, which had a positive effect in a small group of hunter-gatherers, now prompts us to want more than our neighbours. We proudly show off our latest, more advanced car, fashionable kitchen gadgets, bathroom fixtures and several new outfits a year. We are the most extravagant animal on the planet and we copy each other's waste-disposal behaviour (dropping cans and plastic wrappers on the pavement, for example).

Many disasters are now looming over us that no one had predicted and that threaten life on Earth: global warming, rising sea levels, plastic soup in the oceans, the extinction of species and soil pollution. Our primitive brains are inadequately equipped to deal with issues of this nature. The environmental problems our ancestors faced were local and they could be perceived with their own senses. These days, we no longer observe and perceive things personally, so we cannot come up with an adequate response, let alone a solution.

Take climate change, which 99.9 per cent of all scientists agree is happening. There are still American politicians in the United States (even somebody currently in the White House) who, as soon as there is an exceptionally cold snap, deny global warming exists, and who claim that any measures suggested by scientists to mitigate climate change are a waste of money that would be better spent on arms, or bailing out banks or the car industry. Our primitive brain that functions on a day-to-day basis and responds to immediate experiences (such as an extremely cold winter) makes it hard for us to get a sense of what might await us in time to come. We do know that climate change is indeed happening (data has been collected around the world over many generations); a new evolutionary challenge for which we are mentally barely cut out. We 'know' about it, but we do not 'feel' it.

A natural mismatch

Behold the mismatch: we are gradually destroying the planet of our children and grandchildren, because our brain thinks we are still nomads living on the savannah. We do not perceive global environmental problems as our problems. The reality that the ecosystem is threatening to fall apart, with all the terrible consequences for life on the planet, is not registered by our primitive brain.

What gave rise to our present environmental problems? Let's take a look at today's world, using the five ancestral traits of our brain:

1   self-interest first
2   short-term focus
3   the obsession 'to do better' than others

4   copying people around us

5   reacting to immediate sensory cues.

Environmental issues do not feel urgent to us, because our senses are misleading us. Take the problem of climate change. It is more than likely that human behaviour is largely responsible for global temperature fluctuations. Humans are accountable for greenhouse gasses emitted by the food, oil and coal industries. But we do not feel, see or smell those emissions in our own immediate surroundings, so do not take any action. The sky looks clear, our water is drinkable and rain is bound to follow a few days of bright sunshine. Scientists have amassed an overwhelming body of evidence to show that everything is far from OK. Since the industrial revolution, the concentration of the greenhouse gas CO_2 has increased sharply; the average temperature on the planet has risen, as have ocean temperatures and sea levels. According to a study from 2008, the amount of greenhouse gases in the atmosphere is the highest in 800,000 years. The rise in temperature caused by these gases has led to climate change, which is evident from a whole range of factors. Take a deep breath . . .

Extreme weather is on the increase. On average, there are more extremely hot days and fewer extremely cold days. The number of heatwaves is increasing globally, as are heavy precipitation, tornadoes and hurricanes. Many glaciers, along with the sea ice at the North Pole, are melting. The tree line is shifting, and spring is starting earlier in the year. And that is just the beginning . . .

The use of fossil fuels (particularly for energy-guzzling products like cement and steel) is one of the main causes of this climate change. But intensive farming, the ever-expanding livestock sector and deforestation are helping to turn up the heat as well.

Some people see the answer in nuclear energy, a solution prompted by our Stone Age short-term desire for instant luxury and comfort.

The solution to our massive use of fossil fuels is thought to be nuclear power stations, producing cheap energy with zero carbon emissions. But no one has given serious thought to the consequences of this technology, especially the processing of waste. The problem of radioactive waste is being passed on to subsequent generations. And it would only take one big nuclear disaster (like in Chernobyl or Fukushima) for all the perceived advantages to melt away like snow zapped by gamma rays.

Competing for status

It does not feel to us there is a shortage of natural resources, because everything seems inexhaustible. Our brain tells us to keep on consuming, so that's what we do. We buy things to keep up in the competition for status and copy the extravagant behaviour of others around us, instead of just buying what we really need (creating mountains of waste and depleting natural resources in the process). Competing for status is implanted in us by evolution. That has left us with a mismatch, in that this trait influences our behaviour to the detriment of the planet. We receive insufficient cues from our environment to signal that resources are running out. We read about it in the papers and see something on TV, but our impression is that it does not directly affect us. We act as if the resources we use to produce and consume things are inexhaustible.

But forests are disappearing, soil and habitats are drying out, and becoming more acid and arid as a result of our consumption patterns. Excessive use of fertilisers is a problem, for instance, flooding surface water with nitrates and phosphates which leach into the soil, causing one particular plant species (particularly algae or higher water plants) to dominate over other species in a process called eutrophication. Basically, the flood of nutrients allows certain plants (typically algae) to flourish at the expense of other species, disrupting the normal functioning of the ecosystem. If you look down on Earth from space you will see a planet that appears to have enough land for the world's population. But that's only at first glance. The distribution of land around the globe is not equal: at the poles and very high latitudes it is not suitable for supporting a wide range of flora and fauna, while around the equator there are large arid deserts. Soil impoverishment is something that has always happened, even while humans still roamed one part of Africa in small groups. Now, thanks to the spectacular increase in the world's population and mass consumption, we are contributing to it on a vast scale.

Impoverishment means the soil in entire regions becoming depleted and infertile, either temporarily or for good. We are producing too much in too short a space of time and are pumping up too much water, causing the groundwater level in many areas to sink to such a degree that the land dries out. This will leave us with smaller and smaller parcels of fertile land, which will be coveted by increasing numbers of countries and nations. Less affluent countries are – quite rightly – beginning to claim their share of the cake, resulting in ever more intensive use of available land.

The dodo's fate

We have altered the natural environment radically because of our ancestral needs (self-interest, impulsiveness, status). Thus far we have concentrated on the environmental damage we have been inflicting on ourselves with our primitive brain. That is just one side of the story. Because of our behaviour we are also creating various mismatches for other life forms on Earth, at times with dramatic outcomes.

Despite the notion of the ecological savage, man has always had an obvious impact on the survival of larger animal species in particular. Over the course of the centuries, our ancestors have caught and devoured many animals. In 1973, Paul Martin from the University of Arizona discovered a causal link between the rise of

Homo sapiens in certain regions and the disappearances of animals weighing more than forty kilos. He formulated the ancestral overkill hypothesis which presupposes that people – and not changing geographical or climatological circumstances – have been responsible for the extinction of animal species such as the mammoth and woolly rhinoceros. The giant ape *Gigantopithecus* – three metres tall, 540 kilos dressed weight – was no match for *Homo ergaster* in China. The short-faced bear from North America and the Giant Jaguar from South America vanished as soon as our ancestors crossed the Bering Strait. And yet the impact of ancestral humans on the well-being of flora and fauna has been very small compared to that of modern people.

Becoming extinct is an extremely harsh way to die, Dutch writer Koos van Zomeren once remarked. There are really two ways of becoming extinct: in the first one a species disappears from the planet in its entirety, and in the second a species dies out in the wild. Some animal species are only found in human captivity. Biogeographers (scientists who want to know why some species occur in one place and not in another) use the rather bureaucratic term 'functionally extinct.'

The question is: what is the impact of humans on the disappearance of animal species? Extinction is something that has always happened. Most species died from illness, predator attacks or because changes in their living climate dramatically altered the flora and fauna around them. Scientists suspect that over 99 per cent of all species that have ever lived on Earth have disappeared. It is usually alleged that most species have 'a shelf life' of around ten million years. This would mean that the taxonomic animal species called *Homo sapiens* would have quite a while to go yet, because we have only been around a hundred thousand years or so. Still, recent developments on Earth suggest that humans' time on the planet might be considerably shorter than ten million years.

In 1998, four hundred leading biologists were asked their thoughts about the extinction of animals. The vast majority thought we were in the early stages of a massive wave of extinction. Seventy per cent of these biologists believed that 20 per cent of the planet's present animal species will be (functionally) extinct within thirty years. That is a grim thought. Another doom-laden forecast came from the famous biologist Edward Wilson, who predicted that within a hundred years, half of the Earth's current species will have disappeared as a result of human action.

In his beautiful book *The Song of the Dodo*, writer David Quammen describes how this process can happen. Biological interest once brought Quammen to the island of Guam, where an ecological disaster was unfolding at the time: in a short space of time, half of all the island's bird species died out. The investigation into the cause of the disaster lasted a long time, because the ecosystem proved to be hardly affected at all. In the end, the culprit turned out to be a species of snake that had found its way on to the island forty years previously. Quammen was struck by the fact that one species alone could be responsible for the extinction of large populations of birds. He began to look for books about fragile ecosystems on islands, but as he was not able to find any, he decided to write one himself, a catalogue of how and why animals on islands become extinct.

'Let's start by imagining a fine Persian carpet and a hunting knife,' Quammen opens his account. He orders us to cut the rug into thirty-six pieces and place these next to each other. 'Have we got thirty-six nice Persian throw rugs?' he wonders. 'No. All we're left with is three dozen ragged fragments, each one worthless and commencing to come apart.'

This metaphor is *The Song of the Dodo*'s essence. The once so rich and flourishing continent is now so fragmented that the tiny patches of natural environment appear to behave like islands. It becomes a collection of these isolated islands and as the concepts 'extinction' and 'island' are inextricably bound, Quammen concludes that, in this increasing fragmentation, the animal species on main landmasses are also doomed to disappear. The human species has set in motion an all-out flora and fauna Final Solution, which seems to be inevitable.

The gist of Quammen's tale is that islands shed light on evolution, and that, paradoxically and more simply, the biogeography of islands is more peculiar than the biogeography of mainlands. Island biogeography, Quammen explains, is full of cheap thrills. 'Many of the world's gaudiest life forms, both plant and animal, occur on islands. There are giants, dwarfs, crossover artists, nonconformists of every sort.'

This is because islands, due to their isolation, lack many mainland predators and because the animal species need to compete less amongst themselves. This results in 'ecological naïveté': as animal species have never had to guard themselves against enemies, they have developed into the most extraordinary creatures. They are also very easily rendered extinct, having lost their fear of predators and chefs over time.

The dodo of Mauritius is perhaps the best-known example of an animal that hopped around quite happily until it allowed itself to be annihilated by humans. The animal was discovered by a Dutch expedition in 1598, and was initially called the *walckvögel*, or disgusting bird. The dodo proved to be an extremely useful source of fresh meat for travellers on the Indian Ocean. Not devoid of sarcasm Quammen notes: 'Unlike the giant tortoises, they weren't stockpiled alive on shipboards, passing weeks and months in a state of stoical dormancy. Instead, when the supply was excessive, some dodos were salted or smoked; others were eaten fresh [. . .] We can imagine the shipboard menu – boiled dodo, roast dodo, pickled dodo, kippered dodo, dodo hash.' Sixty-four years after their discovery the last living specimen was seen. Led by the Dutch and later the English, all the animals were killed.

What makes the dodo special is that its extinction marked the first time in the history of humanity that humans realised they had caused the extinction of a species. The dodo is the Adam amongst the animal species, wiped out by humans. Quammen cites a biologist, saying that the extinction of the dodo was a supremely important moment for the awakening of human consciousness.

And then he imagines the last, lonely surviving dodo.

Imagine this fugitive as a female. She would have been bulky and flightless and befuddled – but resourceful enough to have escaped and endured when other birds didn't. Or else she was lucky [. . .] Imagine that her last hatchling had been snarfed by a feral pig. That her last fertile egg had been eaten by a monkey. That her mate was dead, clubbed by a hungry Dutch sailor [. . .] She no longer ran, she waddled. Lately she was going blind. Her digestive system was balky. In the dark of an early morning in 1667, say, during a rainstorm, she took cover beneath a cold stone ledge at the base of one of the Black River cliffs. She drew her head down against her body, fluffed her feathers for warmth, squinted in patient misery. She waited. She didn't know it, nor did anyone else, but she was the only dodo on Earth. When the storm passed, she never opened her eyes. This is extinction.

Threatened animal species

The American bison (in Latin, *Bison bison*) came within a whisker of the same fate as the dodo. As a consequence of our interference in their living environment – with horses, railway tracks, trains and guns – there are only as few as 30,000 specimens left, spread over three nature reserves. This is in stark contrast to the millions of specimens who travelled from north to south and back again along the so-called bison trails. It is a misconception that is was only the white colonists who massacred the bison; the Native Americans, too, had their way with them – albeit after the arrival of the Europeans. From the beginning of the nineteenth century onwards the animals were systematically butchered. They were seen as parasites who ate the food available for cows. Bison were hunted from all corners, as they had a fine hide which was especially popular in Europe. American railroad companies abhorred bison, as the animals had a habit of coming into involuntary contact with locomotives, resulting in damage and delays. Their inability to avoid trains happened for the same reason that cars claim so many human casualties these days: we do not have any adaptations for machines hurtling towards us. In 1884, the bison was all but extinct. Years later Neil Young would sing: 'One day shots rang across the peaceful valley/God was crying tears that fell like rain/Before the railroad came from Kansas City/And the bullets hit the bison from the train.'

Humans have created mismatches for other animal species in ways other than through hunting (bison) and food consumption (dodo). New technologies can also give rise to mismatch. By controlling light, we have determined the fate of moths. For millions of years, moths have adapted to navigate using moonlight, and for millions of years this worked out splendidly. Until humans came along. People learned to control light; we made campfires at night and later turned on street and terrace lights and stadium lamps. Those poor moths did what nature prompted them to do: they flew towards this radiation, as moths must. The result was that many of them volunteered to be grilled.

Some animal species have managed to save their skins by adapting genetically to a natural environment affected by humans. A fine example of this is the so-called peppered moth, perhaps the most famous moth in biology. There are two major morphs of this night-flying moth: one white with black spots and the other almost entirely black. During the industrial revolution in England, the peppered moth only appeared in its pepper-and-salt incarnation: white with black spots. Just 1 per cent of the butterflies born were entirely black and these creatures had a much lower chance of survival as, like their brothers and sisters, they lived for the most part on the trunks of white birch trees.

And then the factories in the vicinity of these birches began to pump out quantities of soot, turning many of the trees (as well as houses and streets) jet-black. The result was that the survival chances of the black peppered moth increased, while the white variants became much more visible to their natural enemies. Whereas the ratio of white to black moths was 99:1, in urbanised areas this flipped to 1:99. Once new regulations let to a decrease in soot emissions from factories, numbers of black peppered moths fell as well. Proof that human cultural actions have immediate unintended consequences on other animal species. The fate of many animals lies in the hands of humans – and this causes much mismatch and misery.

Trains, cars, lamps, chimneystacks, factories, fertilisation, irrigation, land reclamation, tree-felling, intensive agriculture, trawl nets, pleasure craft... humans have (often unintentionally and in a short space of time) dramatically changed or ended the lives of many animals by creating a new environment to which these animals were not adapted and which their animal brains were not able to handle.

Green cues

What can we do to mitigate the damage we are inflicting on Mother Earth with our primitive brains? We can start by doing nothing. Perhaps things will run their course. Natural selection of human traits responsible for environmental damage, such as short-sightedness, may emerge. Perhaps evolution will select for people with a large degree of self-control (as we saw in the marshmallow experiment), since they are more successful. If this continues, eventually there will only be people with long-term vision, and this will be good for the environment. But we are not holding our breath, because ingrained traits like interest and status sensitivity seem hard to change. Still, there are things we can do to nudge our behaviour in the right direction, as we will see shortly.

Can politics offer a solution? One characteristic of politicians is that their time horizon does not extend beyond the next elections. That is too short for fundamental changes. The failures of the Kyoto and Copenhagen climate conferences are a vivid example of how not to do it, but maybe the international agreements of Paris 2015 will be able to turn round the threatening global problems to some extent, even with the absence of one of the biggest polluters, the USA. There are some positive examples of political intervention, for sure. Take the hole in the ozone layer. Our atmosphere protects all life on Earth by absorbing the sun's harmful radiation. A layer of ozone – a gas formed out of oxygen atoms (its chemical formula is O_3) – ensures that our atmosphere is protected by a film. So-called CFCs (chlorofluorocarbons), or propellants, attacked the ozone layer, especially in areas with extremely low temperatures. In the stratosphere above the Antarctic and the North Pole, holes in the ozone layer appeared. This resulted in harmful UV-radiation being able to reach the Earth.

This problem was discovered in the 1970s, whereupon quick action was of the essence. On 16 September 1987, the international community signed a treaty called the Montreal Protocol, which included measures to protect the ozone layer. The use of propellants and other harmful gases was scaled back swiftly. In 2005, the Royal Meteorological Institute predicted that the ozone layer above the South Pole, the area most affected, will have recovered by around 2050. In other words, the international community is perfectly able to take responsibility when called upon.

Bringing the future closer

In our efforts to combat this calamitous situation we should consider the traits that helped us survive on the savannah for two million years. To recap (and reiteration will definitely reinforce the message in this case): our ancestors focused on the here and now, their first priority was looking after themselves and their families, they competed for status, copied others and were led by immediate sensory perceptions.

We must not allow ourselves to be overwhelmed by the enormous impact of global environmental problems, but look instead for small, personal solutions. Living in the here and now, we need to bring the consequences of current environmental and climate issues much closer to home. Al Gore opened many people's eyes with his film about global warming *An Inconvenient Truth* (2006). In it, he visualised the steep rise in temperature and the concentration of CO_2 in the atmosphere by going up in a cherry picker. He showed that since the beginning of the industrial revolution, the level of harmful greenhouse gases has been rocketing upwards in one straight line. If nothing changes, the cherry picker will shoot up several metres higher over the coming years.

Visualisation of this kind is very useful. It is already used in public health campaigns. A good example is the computer programs which show what your face and body will look like in twenty years' time if you carry on smoking or drinking. Not only films, but computer animations and 3-D programmes could bring home what the world will look like if half the animal and plant species die out, or if large volumes of fresh water are polluted – and what the impact will be for us. One single image of an empty water reservoir near our house is much more effective than a hundred images of dried-up rivers in China or Africa.

How could we use our self-interest as a means to prevent mismatch? Evolutionary theory teaches us that humans are guided by genetic self-interest. This means they are concerned about the welfare of their relatives and offspring. So if you want to get people to put solar panels on their roofs you might persuade them by pointing out the benefits for their children and grandchildren. They will live in a better world. Studies show that this works. If people were asked to do something that benefited the environment, they would be more likely to do so if the positive effect for their children, rather than the benefit for humanity as a whole, was accentuated.

Deploying role models that resemble you is also effective. People are more prepared to do something for somebody who looks like them, because that is a sign of kinship. Could computers be used to link people up with avatars that resembled them, to ask them to do something for the environment? It sounds like science fiction, but nothing is impossible thanks to a special virtual reality headset called Oculus Rift, launched in 2016. Closer to home, you could get neighbours to raise funds for environmental causes. Studies show that if someone you know, or at least recognise, shows up at your door you are much more likely to give generously.

We could also exploit the reciprocity principle – the idea that one good turn deserves another, which we encountered earlier. It is a form of self-interest and is already being applied with some success in the re-use of towels in hotels. If hotel guests are asked to re-use their towel in the interests of the environment, only a handful do so. But if they are told that the hotel donates a certain amount to charity, e.g. the WWF, for every reused towel, far more people participate. Even more effective is being told that most guests are already reusing their towels.

Green is sexy

Our status instinct is another booster of larger global environmental problems, unless we are able to turn matters round by awarding more status to sustainable behaviour. Many people like to emulate role models like Leonardo DiCaprio. So when DiCaprio and other

Hollywood stars such as Cameron Diaz, Harrison Ford, Will Ferrell, Robin Williams, Tom Hanks and Kate Hudson were photographed driving a hybrid Toyota Prius, the reputation of this expensive but environmentally-friendly set of wheels shot up. Fame sells. Both in America and in Europe it was declared Car of Year. Something similar seems to be happening to the electric Tesla, a car that is also gaining in popularity across Europe, despite a price tag of around 70,000 euros. Business people like to be seen in it. It gives off a powerful message: I am rich, but I care. An ideal peacock's tail.

Daily, many celebrities are deployed for good initiatives to do with the environment. The internet site Looktothestars keeps a tab of which stars are involved with which good causes. No less than eight hundred international celebrities have dedicated themselves to 181 international environmental organisations such as Oceana (including Nicholas Cage and Glenn Close), the United Nations Foundation (DJ David Guatta), the Clinton Foundation (Kevin Spacey and Oprah), Greenpeace (Madonna, Pink and REM's Michael Stipe), Global Green, WildAid, etc., etc.

There are other ways to give green products more status. This can – paradoxically – be achieved by raising their price, because this raises their status. Researchers asked people to choose between two versions of one product, one environmentally-friendly cheaper version, and one environmentally-friendly more expensive version (for instance a washing machine or a car). Respondents on average had a preference for the cheaper version, except for when they had been made 'status oriented' first. Methods to achieve this exist; researchers can tell the test subjects that they are in contention for an attractive job. Sexual motives, too, can make men status oriented. If a heterosexual male had seen a series of attractive female faces, he would choose the more expensive, green product far more often. Our message is: green has to become sexier. Sex sells. We could launch a high-priced racing bike as an alternative to a convertible for men to ride around the streets on to attract women's attention. Sunday afternoon in Surrey would look quite different! We could call this intervention – the sale of expensive, attractive environmentally-friendly products – the Leatherhead effect. One of the most expensive racing bikes in the world is Italian, the Cipollini CB1000, named after the former Italian sprinting champion. The bicycle is yours for 53,000 euros. But then again it is of exceptional quality. The logo alone is made up of forty grams of 18 carat gold, twelve grams of platinum and 17 carat diamonds.

Another idea is the introduction of 'green dating sites'. An environmental site such as Treehugger could include a dating option where people could meet each other for conversations about nature and love. We know from studies that women find a physically fit man more attractive. Men who present themselves on such sites as energetic cyclists or recyclers rather than as drivers of gas-guzzling SUVs stand a greater chance. The Netherlands has the green dating site Crusj, which offers sustainable holiday trips for highly-educated singles. According to its website, the accommodation's sheets and towels are not changed daily, and the air conditioning and lighting are only activated when the guests are in their rooms. The singles also have to pay a premium on their flights to compensate for their CO_2 emissions. The proceeds go to GreenSeat, an organisation that invests in sustainable energy projects.

We should also encourage companies to make sustainable investments on status grounds. The greener the investment, the sexier the company. All kinds of league tables already exist in which companies are ranked according to sustainability (take the Dow Jones Sustainability Indices, for instance, or Greenpeace's annual table of green energy suppliers). Lurking here is the danger of greenwashing, however (see the recent VW emissions-scandal, for example). Companies and authorities tend to show themselves in a greener light than they actually should.

Green standards

Because people unconsciously copy what others around them do, social standards need to be altered. If this does not happen, the problems will only increase. Psychologist Bob Cialdini described a problem in the Arizona National Park whereby a great deal of petrified wood was being stolen. The park issued the following announcement: 'Your heritage is being vandalised every day by theft losses of petrified wood.' As a result of this notice, theft of petrified wood rose by 300 per cent. Why? Because people thought it was normal to steal it, since many people were obviously already doing so. Before petrified wood was becoming truly scarce, people wanted to take a piece home for themselves (again an example of the 'tragedy of the commons'). A better method is to show that you are in a minority with your bad behaviour. We can convince dog owners to clear up the poo of their walking shit factories by impressing upon them that a small minority of dog owners exhibits this behaviour. If this does not have any effect, the law of large numbers can be deployed. It is better to say that as many as 30,000 people are already carpooling on the M1, than to broadcast that only 12 per cent do so. Our Stone-Age brain is not able to distinguish between very large numbers. The difference between 30,000 or 300,000 does not mean very much to us. Many traditional cultures count as follows: 1, 2, many.

The American customer engagement platform OPOWER has come up with an interesting innovation to reduce electricity use, whereby social standards are used. In their electricity bills, the company not only gives information about the account holder's own use, but also about the average use in the area. That way, customers can see if they score above or below average. In addition, the company gives people a smiley if they score better (read: use less than average) and a frowney if they end up above average. This appears to help. People who find out that they use more energy than the average became more frugal at a stroke. These measures ended up providing an energy reduction equal to the electricity use of 150,000 houses. Neighbourhood Power is a local initiative by Dutch energy company Enexis in conjunction with residents' associations. Local people are able to compare their energy bills via a website. It is expected that it will encourage people to save energy. A similar trial in Camden, North London, in 2013 showed a 6 per cent saving in gas and electricity bills. Yet the energy companies will have to surpass themselves to support these initiatives, because there is a mismatch. Financially, it is more attractive for these companies if households use more energy. For that reason, the government will have to intervene.

In your face

If, in the past, there were environmental problems, we would see and smell these. We did not want to pollute our own living environment, which is why we defecated in separate caves (and the Romans much later manufactured their rancid fish sauce called *garum* far away from the city). We move our problems elsewhere, because the problem is then outside our field of vision and outside the reach of our noses, ears and eyes. The sea beds of the Pacific and Atlantic Oceans are littered with containers of nuclear waste from the Netherlands, Belgium, the UK, and US. This is a mismatch: because these problems are outside our field of vision, our brain is not able to respond to it and we will do nothing to sort it out. Out of sight, out of mind. How can we prevent this? It is not beyond clever marketeers to come up with ways to make environmental problems more tangible. An example might be to add an unpleasant smell to things that are bad for the environment, like unleaded petrol. When we see people in certain parts of China walking around with breathing masks to combat smog we think: something is not quite right. An artificial odour has been added to gas coming out of the hob to alert people that the gas is turned on. Why should we not be able do this for urgent environmental problems? We could add a substance to drinking water to show whether it is clean or polluted. The more polluted, the more red. Our primitive brain responds to concrete cues, after all.

Can we do something with our ingrained partiality for nature, biophilia? As mentioned before, nature cues are good for our health. But do they make people more environmentally friendly? In a study we asked people to look at pictures of nature or of a city. We then asked them to take a few financial decisions. For example: you will receive ten euros now or fifteen euros in ten days' time. The outcome? People who had viewed natural images tended to go for the long-term reward. City viewers were more impulsive. Being in direct contact with nature can promote self-control, and this is what we need to safeguard the environment from total collapse. This is something we could foster in children. We are involved in a project into the effect of 'greening' school playgrounds and its impact on the children's social, emotional and intellectual development. The initial results show that boys are better able to concentrate after a break on a green playground, while the grey playgrounds seem to work better for girls. To be continued!

CHAPTER 9

Virtual reality

The fourth of July 2006 is a swelteringly hot day. Bus driver Helmut Haussmann is far from happy with this weather. It is far too close for him. This morning, Helmut has taken a walk in the centre of his hometown of Fürstenfeldbruck, near Munich in Bavaria. He has bought a few newspapers to read up about the match this evening at nine at the Westfalen Stadium in Dortmund. Germany has to beat Italy to get to the final of the Soccer World Cup. Helmut is extremely excited about the prospect of winning the Cup at home; and the whole of Germany with him. Apart from Helmut's wife, that is. Football leaves Lotti Haussmann stone-cold; more than that, it annoys her how her husband can completely lose himself in it, and the whole of Germany with him. It is all that everyone, Helmut included, is talking about.

On a café terrace, Helmut reads about the preparations and chats to a few passers-by. The entire nation is obsessed with the World Cup and everyone assumes *die Mannschaft* will easily beat the *Azzurri*. The omens are good: no German team has ever been beaten in Dortmund. Someone utters the term *Spaghettifressers* (Spaghetti eaters).

Back home Helmut browses the internet for information about the game. He watches clips and places an online bet for twenty-five euros on his country winning. The amount he will receive will be negligible, but it is the idea that counts.

Just as her husband's behaviour irritates Lotti, she also irritates him. She makes increasingly cynical remarks. Helmut has bought a CD-single of the song *'54, '74, '90, 2006* by the German Indie-band Sportfreunde Stiller, who hail from a neighbouring town. Helmut has never bought a single before, let alone one by an alternative band. The song, which depicts a future in which, after 1954, 1974 and 1990, the Germans will once again become World Champions in 2006, is number one in the music charts this week.

When he plays the number for the sixth time in a row, Lotti turns off the CD-player.

'And what if they don't win, Dumbo?'

Meanwhile, thirty kilometres down the road, a team of medical scientists at the University of Munich are deeply involved in the preparations for a study they will be supervising that evening. They have done their homework and have gone over the data from 2003 and 2004 with a fine-tooth comb. During the past few months, links have been made with fifteen large A&E hospitals in Bavaria. Other researchers, including scientists from the University of Utrecht Medical School in the Netherlands, have done similar research in previous years, by delving into the archives retrospectively. The Germans have decided to take a more rigorous approach and conduct their investigation live. Live, meaning during the game.

Helmut is too nervous to eat. Lotti has prepared some tasty food for him, but he is literally unable to swallow. He is oblivious to what she has served up.

When the match kicks off, Helmut's spirits are still high. While Lotti pretends to read a book, but is actually sneaking a glance at the game, Helmut relentlessly spurs on the players of *die Mannschaft* through the TV-screen.

'*Deutschland, Deutschland!*' he shouts, solo. At times, the tension is literally too much for him.

'They're just a bunch of pizza boys!' he screams as the Italians advance towards the German goal. As the match progresses the 'pizza delivery boys' prove to be more in the game; they have more possession and get to the German's goal line more often.

Meanwhile the scientists in Munich remain in contact with the fifteen Bavarian clinics. Information is transferred, the first data is analysed. Even though it is still early, a distinct trend seems to be becoming apparent.

At the end of normal play the score is still 0–0, to Helmut's dismay; he had hoped the enemy would have been caught napping. It gets to extra time and the chances bounce back and forth. Helmut, who has not eaten anything all day, tells Lotti he is beginning to feel hungry. And then, in the 118th minute, the Italian star player Pirlo threads a brilliant no-look pass to Fabio Grosso, who scores the opening goal from a deep position. The whole of Germany falls silent. Just as the Italian forwards manage to net their second goal, Lotti serves her husband a plate of food. Dejected, he looks at what's on his plate: spaghetti. He turns red.

Twenty minutes later Helmut is taken by ambulance to the A&E Department of Fürstenfeldbruck Clinic for emergency heart treatment. Less than an hour later his admission is being processed in the German scientists' study. When the results are published in 2008, the conclusion is shocking. For German football fans over the age of forty-five the risk of a heart attack increases on average by a factor of 2.66 during German international matches. Normally, just under fifteen people a day suffer a heart attack in Bavaria, but on the days *die Mannschaft* played, the University of Munich study revealed the average was a staggering forty-three, all men. Researcher Gerhard Steinbeck pulled no punches in the *New England Journal of Medicine*: 'Viewing a stressful soccer match more than doubles the risk of an acute cardiovascular event.'

Helmut Haussmann survived the World Cup semi-final. He now

takes a beta-blocker before an exciting game. Lotti continues to cook pasta for him at regular intervals. She still calls him Dumbo.

Media

According to the OED, media are 'the main means of mass communication'. Wikipedia groups media under 'communication' and defines this as 'tools used to store information or data'. This can be further differentiated into the technique used (print, digital) and the sensory form of the information (auditory, visual). By means of communication we mean the whole gamut, from smoke signals, spoken language, cave art, clay tablets, legislation, stories and books to photography, film, TV, internet and mobile telephones.

Over the course of human evolution, communication about direct experiences ('Did you see that big lion approaching the camp yesterday?') has been largely replaced by communication about virtual experiences ('Did you see the documentary about that big lion yesterday?'). There is a real world, a place in which we burn our fingers if we hold them too close to the fire and bang our heads against low-hanging branches. But there is also a virtual world, where the imagination is in charge and things happen that are not real. Our actual world, the one in which we live, is becoming increasingly virtual. We constantly read, hear and see things that have little if any impact on our daily lives. The films we watch are not true stories, the books we read are fictitious and the news items we follow are about events that happen in faraway places. And we follow each other on social media, without meeting face to face as our ancestors did.

Living in a virtual environment creates all kinds of mismatches, with potentially harmful consequences for our physical and spiritual health. Our imagination is so powerful it constantly misleads our primitive brains, and vice versa. The influence of new media like radio, TV and internet is sometimes so great that it affects our mental and physical health and can even shorten our lifespan. The question is how to turn this mismatch into a match. Better still, can we harness the power of the media to increase our well-being and solve real problems (like climate change)? That's what this chapter is about.

Tales from the Stone Age

Earlier, we gave a rough sketch of how people lived in ancestral times. Many modern means of information transfer did not exist yet, so this might lead us to think media did not exist in prehistory. Yet this would be incorrect: our ancestors themselves were the medium. People absorbed information that could be perceived directly via the senses. If something was rustling in the undergrowth and one of our ancestors witnessed this, his brain would respond instantly. There was no virtual reality: everything our ancestors smelled, heard, felt and saw, was a direct signal or a real threat. Our entire system was (and is) geared to detecting what the dangers were in our environment; to what was new; and to what opportunities there were for eating, drinking or sex. Our sensory systems were permanently focused on increasing our chances of survival and ensuring that danger or risks were averted. In evolutionary psychology this is called error management theory. We always tend towards caution. If we hear rustling in the undergrowth and we see something move, our first thought is a serpent, not a twig. It increases our chances of survival to assume the worst ('it's a serpent'), because it would be costly to assume it is a twig when we are actually being cornered by a serpent. Error management (or EM, in scientific parlance) recognises two types of errors. Type I is when a suspected snake turns out not to be a snake, but a rustling leaf; or when a judge finds an innocent person guilty in a criminal case. A type II error is when a rustling leaf proves to be not a rustling leaf, but a poisonous snake; or when a judge acquits the guilty person. Errors of the first type are annoying (especially when we are dealing with an innocent person being convicted), but not harmful. Type II errors can be potentially fatal (especially when it concerns the acquittal of a psychopath who will be able to strike again). We can safely state that humans have a brain that is hypersensitive to danger.

But how useful is this mechanism in modern times? EM causes our response to dangers reaching us via modern media to be extreme, even though we are barely in direct danger ourselves. When, thousands of miles away, two planes fly into a skyscraper, our alarm system immediately goes on high alert. We are glued to the box and crave information about perpetrators, victims, motives, etc. It keeps us awake at night. 9/11 has become a 'flashbulb memory': everyone knows where they were at the time. Our primeval brain errs on the side of caution and assumes that a terror act of this kind can happen at any point in our city as well. How real is this danger? Should you forget about that flight to Rome? Do you avoid tall buildings?

In prehistory, we were direct witnesses to events, because we saw them with our own eyes or heard them with our own ears. Until 'language' made its appearance. This enabled our ancestors to pass on information and to become indirect witnesses to events. And so a tribe member would be able to tell one of our ancestors that he had seen the bashed skull of a fellow hunter somewhere or other. This poor chap had obviously been killed as a result of violence. That meant danger! When an expedition was launched to this somewhere or other, they had to bear in mind that there were killers around. It might even be better to eliminate enemies preventatively, so a raid on a neighbouring tribe would be launched without delay.

Earlier in this book we argued that scientists assume that our ancestors learned to communicate using language one hundred thousand years ago, but in fact this is a hard-fought topic of debate, because these manifold scientists do not agree whether we began to learn to talk two hundred, one hundred or sixty thousand years ago. Besides, the question is whether language is innate (nature) or learned (nurture). The nature camp seems to have won, because although language is an extremely complex skill, throughout the world children learn to talk more or less effortlessly at approximately the same age.

The British evolutionary archaeologist Steven Mithen argues in his book *After the Ice* (2003) that human history made a huge leap forwards shortly after the appearance of *Homo sapiens* fifty thousand years ago. Language took flight and humans began to carve images, string beads, paint in caves and make rudimentary musical instruments. This era is called 'the creative explosion'. It occurred in various parts of the world at roughly the same time, which suggests something special was up with *Homo sapiens*. During this period, the 'symbolic brain' is said to have emerged, the typically human quality of being able to create a world outside the known world using language and imagination.

The oldest written sources are a mere five thousand years old, but spoken language is obviously much older. In 1977, a study was conducted about the state of research into the origin of language. There were no less than twenty-three key theories about the origin of language. During the past decades, entire shelffuls of scientific articles about the subjects have appeared, in such quantities that many people are unable to see the wood for the trees. The science that attempts to explain the genesis of language is called 'glottogony' (from the Greek *glotta* or tongue, language, and the word *goné* or generation). This branch of science is relatively young and extremely speculative. We can think up a scenario of how language may have evolved, in the knowledge that it probably happened altogether quite differently. All animals make sounds to communicate with each other. Horses neigh, dogs growl, and birds sing. This stylised emission of animal sound contains a transfer of meaning. Animals can cry out in alarm, chase each other away with imposing noises (it is better to growl than to fight) or court each other: the parakeet with the most beautiful song is more likely to have better genes than his out-of-tune cousin. People are no different.

That our sound system developed into the ability to produce well-turned phrases may have happened in many different ways; either language was incredibly simple at first slowly becoming more and more refined; or language appeared quite abruptly in a complex version, whereupon order was introduced gradually. Whatever the case may be, language is an evolved trait in humans. Like their fellow animals, our language-free ancestors will have uttered frequent monosyllabic shouts and shrieks when they saw a predator approaching, for instance, or hit their thumbs with a celt. Perhaps, certain changing physical and mental circumstances allowed them to acquire the ability to mimic sounds. Onomatopoeias were probably the first 'words' our ancestors were able to utter. A branch snapped off a tree and the word *kchkchkch* was born. *Meh, meh*, meant there was a wild goat nearby. After many centuries and with the expansion of the human intellect, people managed to link various onomatopoeias using a rudimentary form of syntax. *Oe* (you) *ñoe* (gnu) *knak* (neck). This is (obviously) an entirely fictitious sentence for: 'Go and hit that gnu in his neck.'

Language became a social tool for exchanging information, for gossiping. Thanks to language, information transfer expanded exponentially in prehistory. Important events and developments could be conveyed to others. Unlike sharing meat, sharing information was something that could easily proliferate and spread over a large area. David told Hannah something, who could pass it on to Paul and Masha, who went on to tell Jim and Will.

But components could be added to this information, too. Using language, entire new worlds could be created. Whenever it was that language appeared, steps were taken towards a virtual world in the process. This gave rise to possibly the most important invention in human history: imagination. The ability to say one thing and mean something else. It is highly likely that ancestral humans spent night upon night, year upon year, century upon century entertaining each other around the campfire with stories and gossip (see also Chapter

281

4). Language became the most powerful tool to develop a potent group culture.

The singing ape

Along with language came singing (although it may have happened the other way round, singing preceding language). During prehistory our ancestors discovered the joy of rhythm and melody. Important events were accompanied by song and were sung about. Song became a medium for understanding and imparting information about the physical and social environment. The significant questions were always: who are we, where do we come from, where are we going to, what is our identity. These singing sessions took place around the campfire, which – as we saw earlier – occupied a very important place in ancestral life. Whilst people sang, there was movement, dancing, jumping and running. The message of the song was coupled to movement.

In his 1987 book *Songlines*, Bruce Chatwin (1940–1989), explains how the Australian Aboriginals managed to hold their ground in their living environment using song. The Australian land is crisscrossed by thousands of invisible paths which Aboriginals call 'the footsteps of our ancestors'. Songs not only help them to find their way through extensive uninhabited territories, but also to recognise other tribes and their rituals through their centuries-old chants. Chatwin describes how the original inhabitants sang walking from place to place and in so doing, named the world around them. Every place had its own 'songlines' and sacred grounds. For many hunter-gatherer nations, song is still an important means for conveying information and strengthening group bonding. Song and music probably evolved to foster the coordination between people within a group and thus reinforce group culture. Take singing the National Anthem at important occasions, or the various football club anthems thanks to which fans recognise each other, such as

Liverpool's *You'll Never Walk Alone*. In times of stress, crisis or war, music creates a stirring emotion of connection and togetherness. On 12 September 2001, the American Congress joined on the steps of the Capitol to sing *God Bless America*. Nothing brings us together and comforts us as much as music. We are the Singing Ape.

Language as a peacock's tail

The survival advantages that language offers, such as information exchange about dangers and particular people and enhancing social cohesion and group identity, are self-evident, but evolutionary psychologist Geoffrey Miller believes that in the debate about the evolution of language, one function has been left unexplored: sexual attraction. In his thought-provoking book *The Mating Mind* (2000), he argues that verbal courtship plays a central part in human sexual selection. 'Although people may be physically attracted before a word is spoken, even the most ardent suitors will offer at least a few minutes of verbal intercourse before seeking physical intercourse.'

As we saw in Chapter 3, love at first sight is a fuzzy notion, because there has to be some talking before love hits. As a rule, potential love-at-first-sight partners need just under one hour and around ten thousand words (if they talk at the average rate of three words a second) for lightning to strike. Women probably have more need for a substantive exchange of words than men before they fall in love. Miller believes eloquence can be seen as the man's peacock's tail, whereby men use words to show women how clever and cultivated they are, and how much status and power they have. According to this theory, language would be the result of sexual selection on the preference of women for (male) linguistic artists. In our culture language is greatly appreciated for a reason; and as a result men who do not possess a great of deal attraction in terms of physique, try to become brilliant word-spinners in order to safeguard their genetic future. It has always amazed us that, seemingly ugly writers and stand-up comedians get off with the most beautiful women (we will leave it to you to come up with some salient examples yourself).

In the words of Miller: 'complex human language evolved through male orators competing for social status by speaking eloquently, since high status would give them reproductive advantages. As long as those links held true during human evolution, language could have evolved ever greater complexity.'

Miller worked out that the average couple speak a million words before a child is born. The object of the conversations is to get to know each other, make jokes, thrash out arguments and determine whether someone really is sufficiently suitable to have offspring with. The more articulate someone is, the better he or she will be able to present him or herself.

In 2009, the New Zealand literary scientist Brian Boyd published his controversial book *On the Origin of Stories*, an evolutionary approach to fiction and literature. The central question he asks is why we spend so much time on an activity – reading novels – that does not seem to offer us any direct biological advantage. Boyd's answer: because it made our ancestors more social, empathetic and creative. The question now is whether our present time makes different demands from ancestral times. Is there still any point in telling invented stories?

Boyd answers this by referring to traits that evolved initially to serve a particular function but subsequently acquired an additional, different biological exaptation (as these functions are called in biology). Take our eyes, for instance. We not only see with our eyes, we also show something. We can intimidate others with our gaze, we can ascertain the social cohesion of a group via our sclera, because it tells us which ways people are looking. Language came into existence to transfer information. It then became a useful tool for thinking up stories that could strengthen our evolutionary interests. Thus, word-spinners were able to earn high status.

The symbolic mind

Speaking, singing, dancing, music. These were the mechanisms for information transfer in prehistory. A next logical step in evolution was the introduction of art, beginning with body art. Scientists have found evidence that red ochre has been used for more than a hundred thousand years to decorate our bodies with signs and messages. The oldest known cave paintings (at least according to the latest state of research ten minutes before this book went to print) are over forty-two thousand years old, but images were being drawn or painted on walls probably as far back as fifty thousand years ago. Animals were depicted, in both Spanish and French caves (mammoths, aurochs, woolly rhinoceros, horses) and on the Indonesian island of Sulawesi (Babirusa or hog deer). Cave paintings have been found on all continents, except for in Antarctica.

As in all branches of science, many theories exist and many groups of scientists defend their findings tooth and nail. Some believe cave paintings were intended to allay fear of animals, others that they formed part of hunting, dying or fertility rituals, yet another group thinks the paintings were created during a spiritual trance; and then there are scientists who see no more than transfer of information in them. Whatever the case, the paintings must have been produced because people wanted to express themselves, and were able to. And cave paintings are probably the oldest extant demonstrations of self-expression or what psychology calls 'symbolic identity'. So-called hand stencils have been found in many places, both 'positive' variants (hands dipped in paint) and 'negative' (where the hand arrested the paint). In contemporary jargon this would be called a 'hand selfie' or 'handie'!

Our symbolic minds, which mushroomed during the creative explosion fifty thousand years ago, find it incredibly easy to think in highly simplified symbols. Could our traffic signs be distant descendants of these simplistic cave paintings? 'In this area there are giant stags, but beware of bears on the loose!'

Creation stories

Many tribes used language to arrive at a justification for their existence. They constructed a creation history and would pass this on to subsequent generations, something that must have arisen amongst practically all peoples. Wikipedia's list of creation stories amounts to more than a hundred different myths from all corners of the Earth, and this is just the tip of the iceberg. Seen from a scientific perspective, creation myths outdo each other in their bizarreness. Birds trying to push the heavens upwards, ancestors locked up in total darkness, fiery creatures hiding in trees, gods allowing themselves to be fertilised by the strangest objects, supreme beings, giant children, chaos, sex between heaven and Earth, people born out of rocks and emerging from the sea; there is even a creation story in which humanity came into being following a divine act of auto-fellatio.

Thanks to language, imagination arrived on the scene. From direct events we moved to indirect descriptions of events, and needless to say the role of fantasy, as well as cunning and guile, became much greater. Perhaps fuelled by self-interest, people began to invent stories about others that were not true; 'Watch out for Mary, she sleeps around, and has had it off with various men in her former tribe. Joan would be a better woman to fall in love with.'

We all know the game Chinese Whispers. A group of people sit in a circle and someone whispers a phrase into the ear of the person next to him. This person passes it on to his neighbour and so forth until the phrase arrives back with the person who started the game. And then it turns out that this is miles removed from the original phrase. In 2013, a big-scale international version of the game was played. Seven hundred and thirty players from seventeen countries (including Antarctica) whispered a phrase in a minimum of six languages, covering a total of 151,926.9 kilometres in the process. Five branches emerged, and the original phrase 'Play is training for the unexpected' after long peregrinations came back as 'I love the world', 'Zombie', 'Clouds travel around the world', 'Glow, glow, peanut butter jelly' and 'Ian needs help'. Make no mistake: spoken language can be a source of mistakes, deceit and deception.

Agriculture and script

The advent of agriculture led to bigger, more social and more complex societies. Language saw an even greater significant expansion, because an important element was added to spoken language: script. Writing means reading. Writing and reading have been recent developments in our evolutionary history. This means a mismatch is lying in wait. Script has come about through trade, as records need to be kept of all agreements entered into. When a population grew too big, verbal agreements no longer sufficed, and so people began to make a note of transactions. This was done by scratching marks on wooden, bone or earthenware objects.

The Sumerians, who lived around six thousand years ago in present-day south-eastern Iraq, are considered to be the first people to have developed a true script. They carved symbols in clay tablets, a raw material that was readily available in the region. The Sumerians, who called their country Ki-en-gir, or the land of the noble lords, introduced irrigation and their villages expanded into big cities. Clay tablets were used to register laws and business agreements. Archaeologists have found a hundred thousand clay tablets from one of their dynastic periods (from 2100 to 2000 BC) which were used as part of the rulers' administration. In addition to business transactions, Big Stories began to be noted down, historical events, stories about the origin of everything, information about the world, current events. The Sumerians began to number their clay tablets; in other words, a prehistoric form of the book was born.

We have to thank the Egyptians for papyrus, sheets of flattened pith from the papyrus plant suitable for writing on. Other nations (Greeks, Romans) adopted the technique and began to use parchment, or prepared animal skins. Rhetoric, the art of speaking in public, had been perfected meanwhile, and the Greeks in particular became masters at merging the 'oral world' and the 'written world' as two manifestations of the virtual world. Papyrus and parchment had the advantage of being much easier to transport than clay tablets. The Romans introduced the so-called wax tablets, which were wooden tablets with a raised rim onto which beeswax had been smoothed. Using a *stylus* (a piece of wood with a sharp point and tapered top), notes could be made and erased. This greatly democratised writing; every Roman pupil or student had his own wax tablet, just as pupils and students now have their own laptops. Sometimes two or more wax tablets were joined together, creating an ancient form of book.

At the beginning of our era, someone probably came up with the idea to cut parchment scrolls (*biblion* in Greek) into strips and to bind these together. This was the true birth of the book. The pages of parchment were held together by solid leather and square pieces of wood. This is the origin of the word book, which most probably comes from the Germanic for 'beech'.

Nowforanexperimentbecausewewouldliketodemonstratesomethingtrytoreadthissentencewithoutgoingbacktowhatwasatthebeginningofthesentencetheearlystagesofthewrittenbookwerecharacterisedbythelackofseparationbetweenwordsandothergrammaticalconventionsthatmakereadingeasier. That was quite tough; imagine having to read an entire book this way. Initial readers of books were not given an easy time, roughly a thousand years ago, thanks to this 'scriptura continua'. Books were not meant to be snuggled up with in a corner, texts were there to store information or to be declaimed. But by the end of the Middle Ages legibility of texts became increasingly important.

With the advent of printing, in the eleventh century in China and around 1450 in Europe inaugurated by Johannes Gutenberg, book production proliferated compared to the unique copies which had either been laboriously calligraphed or produced using complex techniques. The Chinese had also invented paper, which replaced expensive parchment in Europe. In his 2010 book *The Shallows*, the American writer Nicholas Carr states that scientists have calculated that the number of books produced during the fifty years following Gutenberg's introduction of printing was the same as the number of books written in Europe over the previous thousand years. Books became affordable mass products which appeared in enormous print runs in ever smaller formats. In a relatively short space of time, books changed how the world was viewed. A wave of intellectual energy hit Europe. Carr makes a comparison with the reduction in the size of clocks; thanks to the arrival of watches and small clocks, everyone became a timekeeper. The ubiquity of the book in pocket format led to reading becoming part of everyday life. At the end of the fifteenth century, as many as twelve million books had been dispersed throughout Europe (amongst a population of around sixty million). It was the beginning of the 'information revolution'.

Cultural, religious, scientific, but above all political life developed rapidly as a result of the swift distribution of printed media. Not only did many books appear, but pamphlets also began to leave a mark on the developments to a similarly important degree. A pamphlet is a one-off occasional publication often, but not exclusively, with topical content. During the sixteenth and seventeenth centuries, especially during times of upheaval, the Dutch provinces were swamped with leaflets that took a stand for and against all manner of things. Official notifications, appeals, jokes, anonymous gossip, misinformation, accusations, imputations, mockery, gibes ... pamphlets were the tweets of their day. And the fact that pamphlets were not innocent rags is proven by the story of the murder of Grand Pensionary Johan de Witt (arguably the greatest politician the Netherlands has ever had) and his brother Cornelius. On Saturday 20 August 1672, The Hague was plastered with pamphlets calling for his murder. The inflammatory texts were written by notaries, lawyers, God-fearing clergymen and followers of the Orange family who wanted to assume power. The populace, stirred up by reams of these pamphlets, threw themselves on the brothers in an orgy of blood and violence.

The printed word continued to hold sway until the beginning of the twentieth century. Millions of books, newspapers and magazines were distributed, but the twentieth century produced an entirely different form of communication: electric media. The arrival of photography, telephony, radio, television, computers, internet and mobile communication to a large extent displaced the omnipresence of the written word. It was the Canadian media philosopher Marshall McLuhan who, in 1964 in his controversial work *Understanding Media*, advanced the thesis that the world would never be able to free itself from the stranglehold of technological media.

McLuhan is best known for his phrase 'the medium is the message', by which he meant that the means used to convey a message influence this message. If a particular medium only allows 140 characters to communicate a message, then these 140 characters will partly define the content of the message. According to McLuhan, the content of a particular medium (television, telephone, internet), is 'the juicy piece of meat carried by the burglar to distract the watchdog of the mind.'

The narrative brain

The media offer endless opportunities to realise our evolutionary goals (survive, protect, have sex, acquire status, competition and food). On TV, we watch news programmes, crime series, sports, dating, cookery and health programmes and quiz shows as they provide us with important information. The most searched terms on Google in 2016 were 'Prince' and 'David Bowie' (deceased popstars), 'Olympics' (sport, our sublimation of waging war), 'Hurricane Matthew' (natural disaster), 'Malaysia Airlines' (fear of accidents),

'Pokémon Go', 'Powerball and 'Slither.io' (play), 'Trump', 'Election' and 'Hillary Clinton' (obsession with leaders). By looking at the popularity of our searches on the internet and what we watch on TV, we are able to find out what our ancestors considered important.

We watch crime series to recognise a murderer in our midst. A message in a TV ad embedded in a jingle or poem sticks much better ('Go Compare'). We tell our children fairy tales, whether consciously or not, to alert them to potential dangers in life. The tale of Red Riding Hood teaches children to be wary of strangers and not to be too credulous. Dangers can strike from unexpected corners. Wise lessons for life are also embodied in Sleeping Beauty (stay on guard) and Cinderella (watch out for the wicked step-mother). But what is the evolutionary message of Jack and the Beanstalk?

Evolutionary literary experts such as professor André Lardinois (Radboud University, Nijmegen), and the previously mentioned Brian Boyd, are conducting research into the narrative brain. They are discovering that stories are present in all kinds of different cultures. All around the planet, children and adults are told fairy tales, myths and parables to give them important evolutionary messages. If you study the behaviour of protagonists in novels, as Lardinois has done, you will see that, in the main, it can be reduced to biological motives.

Literature can greatly magnify our primordial motives through the art of overstatement. Stories and novels are invariably about sex, power, violence, fear, desire and conflict. In the novel *Madame Bovary* a doctor's wife commits adultery to escape her torpid provincial existence, an archetypal theme that was so shocking at the time that, in 1857, the writer Flaubert was forced to answer to a judge for his alleged obscenities. The novel *American Psycho*, by the American writer Bret Easton Ellis, contains long and extremely graphic descriptions of atrocities committed on innocent creatures without any moralistic explanation. Two examples of novels that present us with exaggerated clues to which our brain responds acutely, with delight or disgust.

More than a thousand words

Our brain processes the information we receive from various media as if it originates from our immediate living environment and is therefore relevant to us. We did not evolve alongside TV, photography, film or internet. Fake cues of this kind create mismatch. During the millions of years that our human brains evolved there were no photos, as these have only been around for just under two hundred years. The misleading thing about looking at snaps is that our brain thinks the person in the photo is actually with us, until it registers that it is gazing at an image. In her book *On Photography* (1977), Susan Sontag called these fake cues 'pseudo presence'. It is the reason why people carry their loved ones in their wallets; there is a brief recognition, reverie, a reminder of the good things in their lives. It is 'rendering the world immortal'. People change or die, landscapes disappear, buildings fall down, but photographs capture for eternity what once was and never will be again. Recording your daily, familiar environment can be an advantage for someone who goes travelling. Photos induce us to concern ourselves more with people who are not directly in our field of vision. A picture can say more than a thousand words, especially if this image is intended to invoke empathy. In order to get more support for immigrants, relief organisations show images of refugees, preferably women and children, crossing the Mediterranean in dinghies.

Our brain is strongly geared towards sensory experience (something the Russian neurologist Ivan Pavlov described as early as 1927 as the 'orientating reflex'), and in particular towards visual clues. In response to new visual clues we receive brief physical reactions, such as a slowing down in pulse, psychogalvanic skin reflexes with names we cannot even pronounce, vasoconstriction and pupil dilation, and that is why television is such a brilliant medium for grabbing our attention. It is extremely difficult to talk to someone in the presence of a television that is switched on. Our brain is visually focused on what we see and what moves. The phrase 'people look at everything as long as it moves' is true.

As previously mentioned, the threat of danger captures our attention instantly. Even if something that we see on TV or read about in the newspaper has happened far away. Attacks in big cities are immensely fascinating to us, even though we live a long distance from them. We immediately ask ourselves the questions: what is happening in the world? Are we in danger? Should we increase our threat level? Is it a terrorist attack or an accident? Just to be on the safe side, our brain assumes it is an attack (a type I error). In terms of mismatch theory most news reports are inflated or sometimes even obsolete cues to which our brain responds strongly. When, in 1930, Orson Welles reported an alien invasion in a radio adaptation of *The War of the Worlds*, many American listeners actually believed the Martians had landed. In panic they rang the police.

Our brain does not distinguish between real dangers (crossing the road in front of a car) and unreal dangers (flooding in Bangladesh). We seldom calculate our chances based on objective data. Thanks to 24-hours news channels such as Fox, CNN and BBC World we overestimate the chance of being involved in a plane crash, because these are covered widely in the media, whereas, when all is said and done, the chance of having an accident on your bike or in your car on the way to work is many times greater. Take the attack on the World Trade Center on 11 September 2001. The official death count was 2,996 including nineteen terrorists. During the months following, 12 per cent fewer passengers passed through American airports than during the preceding months, whilst the number of road users increased by 20 per cent. German Professor Gerd Gigerenzer, an expert in risk communication, calculated that the Al

Qaida attack resulted in some further 1600 American casualties. This was because people, too afraid to fly, preferred getting in their cars, even for long-distance journeys between cities. Travelling by car is many times more dangerous than travelling by plane. These road casualties were the result of mismatch. As Gigerenzer said in an interview: 'People jumped out of the frying-pan into the fire.' After 9/11, Dutch America expert Maarten van Rossem was asked on TV whether World War III had started. He muttered that he believed it was an incident, an action by a bunch of nutters which would have no effect on the situation in the Netherlands whatsoever. It earned him national derision and anger and the usual hate mail. An attack of this kind can only lead to us thinking 'danger threatens, we have to take action', even if this disaster is occurring on the other side of the world.

The virtual social world is important

Why would you leave your house or bedroom if you are in touch with the rest of the world via television and internet? Psychological research suggests that girls with photos of Justin Bieber on their walls actually fall in love with the squawking pin-up. A British study reveals that people who watch soaps such as *EastEnders* or *Coronation Street* often actively believe the actors are their personal friends. They see these people pop up in their living rooms on a daily basis, after all. Before there was TV, seeing someone in your house every day meant that you knew him or her well. So you sympathise strongly with these people, even though they are actors in a series. Soap addicts can end up isolated and no longer able to maintain real friendships as a result. And, another study tells us, real friendships benefit our mental health. Everyone allows themselves to be tricked, but this self-same British study shows that it is above all people with a below-average IQ who are troubled by this pseudo-presence of TV and film stars.

Regarding Facebook, too, a mismatch is lying in wait. An extensive study from the University of Göteborg into the effects of Facebook use showed that people establish social relations more easily with fellow site users. But frequent Facebook interaction can damage someone's sense of self-worth and mental health. Facebook users generally present their lives as a more rosy picture than they really are, and with more friends. They start to compare themselves to other users who lead an even more exciting and adventurous existence. This leads to a massive status competition, with consequences for their self-image. Women especially indicated that they felt more miserable the longer they spent on Facebook. Other research shows that Facebook is above all a 'self-focused' network. It purports to be social, but there are narcissistic and psychopathological sides to the site. Studies demonstrate that the more narcissistic you are, the more Facebook friends you have and the more information you post about yourself. But it remains to be seen whether you will get help from your Facebook friends when you split up with your partner or when you have run out of money.

Intense internet use and prolonged TV-watching leads to unrestrained status competition. Let's imagine a boy or girl in prehistory. In a tribe of around a hundred to a hundred and fifty members, adolescents had few counterparts to compare themselves with. Some were more beautiful, some less so, some were blessed with more muscles, some with fewer. Thanks to contemporary media overload, boys and girls are able to match themselves against not a hundred but ten thousand others. With one click we can see all manner of models in all manner of sizes. As a consequence, many people are perpetually unhappy about their appearance or the appearance of their partner, because in the virtual world there are always people who are more fun and more beautiful. Researchers showed young adults ten photos of people of the opposite sex who were all either more or less attractive than their present partner. The result? The people who were exposed to attractive alternatives reported they were less happy with their relationship. Today, there is a systematic dissatisfaction with our status and our appearance as a result of a deluge of exaggerated cues. The media cause global competition, whereas in the past competition was resolved in a local battle. This permanent status checking leads to depression and a low sense of self-esteem. Psychiatry and the cosmetics industry are thriving as a result.

Pornography

Over the past two million years our brain has developed into the lively organ we carry around these days. For hundreds of centuries, our brain learned to respond to daily impulses such as a frisking deer in the woods, conflict between different groups or an excited fellow human. Susan Sontag's pseudo presence is a pre-eminent ingredient for mismatch. The sex industry at any rate has reaped profits from it.

Whether we are a man or a woman, when we see a pornographic image, our Stone Age brain thinks: a naked person, I have access to this person and physiologically, I have to prepare for this. In past times, when we saw a voluptuous naked body, it would be in our actual presence. The likelihood that this would result in a sexual encounter was great. As we showed in Chapter 3, sex then, just as it is now, was a private affair.

As sex takes place in seclusion, so does pornography, for the same reasons. Pornography has probably been in existence for as long as our ancestors have been painting in caves. 'Porno' probably stems from the Greek *porne* for 'prostitute', derived from *porneia*, illicit sexual intercourse. 'Graphy' derives from the Greek term for description or illustration. So pornography is literally a text or image of a prostitute or prostitution. These days this description is somewhat limited: in the shops and on the internet billions of films and photos can be found of people who have sex in all possible ways involving all possible orifices and extremities of the human body.

Meanwhile pornography has taken over the world and it is the driving force behind new media technology. Let's take a look at some enlightening figures. It is estimated that 35 per cent of all downloads globally are of a pornographic nature. Or 70 per cent, this is another estimate. At the end of the first decade of this century seventy million pornographic terms were entered into search engines every day, and this constituted 25 per cent of all entries. The word 'sex' was the most-searched erotic search term. Pakistan, India and Egypt saw the greatest number of searches for this word. Croatia and India were the only non-Muslim countries in the top ten of 'sex' searchers. In 2000, an estimated 60 per cent of all website visits were sex-related. In 2006, the number of 'unique visitors' to adult pages was estimated to be seventy-two million a month. The number of sex sites was estimated to be 420 million.

In that same year, the sex industry's global turnover was more than ninety-seven billion dollars, double that of Microsoft for the same period! The Netherlands was at the bottom of the table with 200 million dollars. China had the highest pornographic turnover, almost twenty-eight billion dollars. In South Korea, 525 dollars is spent per person on pornography, a record! America, Brazil and the Netherlands topped the table in regard to porn production. In America, there is a clear correlation between income and porn use. The more someone earns, the more he surfs: people in the highest income brackets (75,000 dollars or more) view 35 per cent of the total. In 2007, a third of all thirteen-year-old-boys in Canada watched dirty images or films. In 2006, 87 per cent of American students had sex using webcams, MSN or telephone. The age group thirty-five to forty-four watched most frequently.

Some more figures. In 2004, 44 per cent of American employees (m/f) admitted having searched for flesh. Amongst university users it was as much as 59 per cent. Seventeen per cent of all women claim to struggle with porn addiction. One in three visitors to sex sites is female. Fifty-nine per cent of adults think it is morally acceptable to have sexual fantasies. Thirty-eight per cent think it is morally acceptable to view pornographic images. Okay, the above figures are not 100 per cent reliable, because the internet is a source of nonsense, overstatement, misinterpretation and malevolence, but one thing is certain: if you go surfing for porn, you do not need to feel alone.

Our question is, where is the mismatch? The answer: the medium is the mismatch. We are at the receiving end of an enormous availability of fake cues to which we have a strong physical reaction. Male brains in particular allow themselves to be manipulated by porn to masturbate instead of copulate. The Dutch for masturbate is *onaneren*, which is derived from the Biblical figure Onan, who aroused the wrath of God because he preferred to ejaculate his seed onto the ground instead of into the womb of his wife. And that's not something you should try on God! In a rage, the Creator struck Onan dead.

These days, masturbating will not kill us, but there are issues signifying a potential mismatch. Sex addiction is a serious problem, because, for example, the amount of time and contributions men 'spill onto the ground' are not invested in begetting offspring. The urge to watch pornography prevents some men from normal daily functioning. Something called SADD-effect (Sexual Attention Deficit Disorder) exists, a sexual syndrome men who watch too much porn or masturbate too frequently are said to suffer from. The effect has been named by the American sex therapist and writer Ian Kerner, who discovered that some (younger) men had become so habituated to 'visual stimulation' and the increasingly intense stimulation of internet pornography, that they were no longer able to turn themselves to a real woman during actual sex. Men who are plagued by SADD tend to have difficulty maintaining an erection during sex. They seem to get bored quickly in bed.

Because pornography is freely and abundantly available via the internet, many men have unwittingly acquired a distorted view of sex. Kerner believes they have developed a typical porn-way of masturbating, which bears only the slightest relation to sex in normal life. Kerner has a simple solution for men who are addicted to pornography: stop watching it for a while! Sexology can be so simple at times. And it may cancel out a possible mismatch in the sphere of sex.

Talking about pornography, a link also exists between love and football. The World Cup finals in Brazil took place a while ago now (in 2014), but some people are still basking in the afterglow. The data controllers of the site pornhub.com, for instance. For those who do not know it, this Canadian site is the biggest 'pornographic video sharing website', the YouTube of sex films. The interesting thing about Pornhub is that it keeps a blog with studies, analyses and striking data about use of the site. Visits to Pornhub are not confined to Canada, but come from across the globe. And when there are global events, such as World Cup Finals for example, remarkable unexpected links can be established.

A water company in the Dutch province of Drenthe reported a while back that during the match between Argentina and the Netherlands, water use plummeted. Half-time showed a sharp rise, because everyone visited the bathroom. The Pornhub research team saw a similar effect. When a country played a game during the World Cup Finals, the number of visitors dropped dramatically compared to other countries. You are not going to watch porn when everyone else is cheering on your country's team in front of the box. In comparison with countries such as Portugal and England, the Netherlands stood out. Orange match days saw the biggest drop in the average number of porn viewers. A match involving the Dutch football team is clearly such a climax, that it renders redundant all other climaxes.

At that particular moment in time. Because after that it's bingo! Following the game between the Netherlands and Brazil the number of visits to Pornhub increased by 40 per cent compared to a normal day. In other words, the Dutch side had made the country significantly more randy. The Belgians went even further: after their victory over the United States visits shot up by close on 100 per cent. The things football is useful for.

Smart, too, was Pornhub's analysis of search terms during the World Cup. It appears many people search using catchwords in their quest for explicit visual pleasure. What turned out to be the case? When a country played against another country, curiosity for the opposition measurably increased, not only amongst the opposition but also in the rest of the world. Pornhub statisticians saw for example that when the Netherlands played Argentina, the search term 'Netherlands' rose by 195 per cent and 'Dutch amateur' by 55 per cent. Football thus unites the world on several fronts.

Some time ago we caught a conversation between two adolescent boys in the city of Nijmegen. They were doing their A levels and were faced with the choice as to what to do next. One of them said: 'The thing I'd like to do most is be a porn actor.'

'Yeah!' the other boy cried, gloating.

Oh well, boys. We may safely assume that the parents of this lad would hardly be thrilled when they heard about his career choice. The profession of porn actor is not particularly well thought of (somewhere between a drug dealer and a crook), whereas – at least in America – a good living can be made from sex in front of the camera. Women earn considerably more in the porn film industry than men but they, too, are perfectly able to support themselves (between 100,000 and 300,000 dollars for women, against 40,000 to 100,000 for men).

Aside from the unpleasant side-effects porn stars encounter during the course of their horizontal career (social isolation, sexual numbing) there is another reason to think a little more deeply about this career move: the knowledge that porn actors can be rather self-destructive. What's ironic is that porn actresses have been programmed to look as fertile as possible and to behave in an as fertile a way as possible, but researchers believe they have far fewer children than the average American, and die much younger.

Out of the 1500 porn stars who were working between 2007 and 2010, fifty-one died a so-called premature death (they contracted AIDS, committed suicide or were killed). By comparison: in 2009 well-nigh a hundred thousand music albums were released against thirteen thousand porn productions. And yet in the music industry, between 2007 and 2009 there were only nine drugs-related deaths and two suicides. In other words, for your health it is better to be a musician than to play in sex films.

A small number of (loathsome) Christian internet sites exist that relish the high death rate amongst porn actors. These sites call it a punishment from God that the average life expectancy of a porn star is a mere 37.4 years, against 78.1 for the average American. Setting aside the fact that statistical holes can undoubtedly be dug in this calculation, it is probably true that porn stars die younger than other people. Many of them come from broken families, have been neglected and have had dealings with (sexual) violence. However attractive the idea of being a porn actor may be to an adolescent boy in Nijmegen, he would be wise to give it a bit more thought.

Copying errors

Our brain is focused on role models and on the question of who within our group has the best qualities to survive on the savannah. Good hunting techniques and strategic insights were eagerly followed and copied for hundreds and thousands of years. People who were held in high regard were much imitated. Hence our modern fixation with celebrities, people we see as good role models in one way or another. If we see a trendy celebrity wear a particular pair of sunglasses, the chances are we too will be sporting them two weeks hence.

Nonetheless, there are some health benefits to worshipping famous people, as psychologist Shira Gabriel from the University of Buffalo has demonstrated. Thinking about someone we admire increases our self-confidence and social skills. A little admiration is healthy, but there are limits. The human brain is, as many studies confirm, not especially well-equipped to differentiate between real relationships and what psychologists call 'parasocial' or 'imagined' relationships. A number of British studies show that the admiration of famous role models can lead to what's known as 'celebrity worship syndrome' whereby people literally become addicted to their admiration, with all the misery this entails.

The fact of the matter is that, these days, role models are no longer individuals who have useful qualities or other people's best interest at heart, and this can have consequences. Moreover, they are able to reach millions with one action on social media, sometimes with disastrous results. In Chapter 1 we discussed the phenomenon of suicide copycats, people who commit suicide because celebrities or other people in their environment have done so. The media could play its part by paying little heed to particular instances, such as the alcoholic female British singer who took her own life (possibly accidentally).

Science speaks of a 'suicide contagion' when there is a 'suicide cluster'. There are reported cases where the suicide of one person 'invites' other suicides. There are examples of suicide waves when a celebrity has taken his own life (this is the Werther-effect, named after the main character in *The Sorrows of Young Werther* by Goethe, a book that is alleged to have provoked a series of suicides, although there is no proof of this). Famous cases are the Chinese actress Ruan Lingyu (1910–1935), the Japanese musician Yukiko Okada (1967–1986) and Marilyn Monroe (1926–1962), whose death is said to have caused more than two hundred suicides. More recently, there was the case of the Tunisian street vendor Mohamed Bouazizi who set himself alight in 2010. Media reports led to people in other countries following suit. In order to stay one step ahead of this kind of copycat behaviour, many countries have laws that prohibit the media from reporting suicides and suicide attempts, Norway and Turkey for example. These laws do not exist in the Netherlands and the UK.

But there are more innocent variants of this media mismatch. School children who say that, instead of learning a trade, they want to be famous, like Kim Kardashian or Joey Essex, without it being at all clear what these people's intrinsic qualities are. Or people who spend a lot of money on cosmetic surgery because their TV idol has done likewise.

Media mismatch can take on more extreme versions. Almost daily, new images of IS acts of war appear on TV and social media. These images stimulate our brain and hardly appear to diminish the attraction of this terror movement to young Muslims. Approximately eight hundred and fifty British jihadists have joined Islamic State to fight in Syria and Iraq. No fewer than one hundred and fifty young men have died, and this high death rate should deter any straight-thinking person from fighting. And yet IS has seen a large influx of young Muslims from Western Europe and elsewhere. Analysts and politicians are racking their brains about the causes of this and proffer explanations related to issues such as discrimination, lack of opportunities, abhorrence of Western culture and the influence of particular imams. We believe there is a much deeper underlying reason. To many young men, waging war is simply an enjoyable activity (see Chapter 7), and a means to boost their status in the community. They see pictures and think: I've got to be part of that, it will make me a hero. To counter this, US authorities have issued an anti-IS video with images of IS organised public executions and suicide bombs going off in mosques. The message is 'you must be mad to want to kill your Muslim brothers', but whether this is hitting home remains to be seen. There are also scientists who suggest it would be better to go to potential IS fighters' direct circle (mothers, influential imams, well-known Muslims) and belabour the point that the people who go to Iraq and Syria are dupes, boys who will be spewed out by their own community. Do not give jihadists the media status they believe they will have or hope to have.

Media and addiction

In order to experience something, we no longer need to leave our homes, because television, the internet (and soon Oculus Rift goggles) offer everything we need in a virtual environment: excitement, sensation, information, involvement with the planet, social contacts with our friends, sports and sex. In earlier times we literally had to get up and go somewhere to witness and take part in adventures and acquire the necessary experience, burning a large number of calories in the process, but this is no longer necessary. There have been many studies into health problems related to watching TV. The general conclusion is that the more time we spend in front of the TV, the more unhealthy we are. An average American household consists of 2.7 people and has an average of 2.9 televisions. This gives an indication of how important the goggle-box is. On average, Americans watch 242 hours per month, and children even more. Add to this twenty-seven hours a month on the internet, with YouTube films being watched the most.

Watching TV for long periods is bad for our minds as well as our bodies. Children who watch a great deal of TV are less proficient at reading and maths and have lower scores in IQ tests. We burn 14.5 times fewer calories watching TV than by simply lying in bed (as a cynic might put it: you are better off watching TV in bed). There is a chicken-and-egg issue in these studies. Perhaps people who are by inclination lazy and stupid are the ones who watch too much TV. But it is obvious that their hobby does not offer them any benefits.

We speak of an addiction when a person is dependent on a substance or habit to such an extent that his or her physical or mental well-being might be compromised. Some people seem to have developed an addiction for TV, email, Facebook or Twitter. They are literally unable to do without and these media control their lives in a way that preys on their own minds. Their obsessive behaviour reduces their productivity and happiness, as a result of which they no longer pursue other, evolutionarily more relevant goals.

Another mismatch problem is that information on the internet is barely filtered – if at all – before it reaches us, resulting in an increased likelihood of us making the wrong decisions. In his book *The Organized Mind*, neuroscientist Daniel Levitin shows how we are submerged in a sea of information. Our brain is not made to store a great deal of information by itself. This means we are in a permanent state of information overload, with unpleasant consequences. Thanks to the wonders of the internet we are now able to do everything ourselves for which we formerly needed specialists: booking travel, publishing photo albums, ordering bicycles, editing films, arranging mortgages, compiling medical information. We end up doing too much, losing too much time in the process and making too many mistakes. Our parents used to have deep respect for their GP, whether justified or not. Our brain operated according to the decision-making rule 'trust someone in a white coat'. On the few occasions someone within the family was unwell, they would call in the help of a doctor or wait until the ailment went away by itself. Abdominal pain, flu, a temperature, a rash in strange places: most complaints would resolve themselves in the end. Only if after a few days you were still feeling rough would a visit to the doctor's surgery ensue. The GP would ask a few questions in a calm tone of voice, would palpitate the odd body part, sit down in silence at his desk to write out a prescription whereupon you could leave the surgery satisfied, almost cured. That is how a visit to the doctor was meant to happen. The doctor's authority was never in question, medical instructions were strictly observed ('complete the entire course').

These days, patients are articulate, critical self-medicators. If someone feels unwell, a whirlwind of googling ensues. Many health and self-medication apps are being developed and the end is nowhere near in sight. People prefer making their diagnoses based on information found on the internet, but this tends to be unfiltered and taken completely out of context. If you make an incorrect diagnosis by yourself and then go on to self-medicate, you run a health risk. Perhaps the best-known example of this was Apple Boss Steve Jobs, who thought he would be able to create his own pancreatic cancer treatment plan. When this turned out less well than he had hoped, he decided to seek treatment via the regular medical channels (which operate according to scientific methods) – it was too late, as his cancer had spread too widely.

Registered in full

In his novel *The Circle*, Dave Eggers paints a near future in which privacy no longer exists, or rather: privacy is theft. The acme of mismatch. The entire world is allowed to know everything about everybody at all times. Life has become one big *24 Hours With . . .* , in which the camera plays the part of host. Full transparency is the aim. This apocalyptic world view is almost a fact. Our smartphones are on all the time, which means that every moment of the day our whereabouts is being registered. Where we park our cars is being recorded, our payments are logged, we take photographs all day long which are stored in a cloud (without it being in any way clear who might be looking at them), throughout the country there are CCTV cameras, apps and cookies keeping track of our internet activities, our mails are accompanied by ads related to the content of our mails, and if you have a quick pee against a tree somewhere you run the risk that some dimwit will snap this and post it on the internet. Women no longer wish to sunbathe topless, for fear that their breasts will appear on porn sites. Lying in the sun in the buff is fine, but not if it ends up on the internet. All this is a mismatch. In our ancestors' small communities people knew a great deal about each other and someone's reputation was forever being tested. But in order to keep the group together and to avoid conflicts, people had to give each other space, and a person's privacy and autonomy were tremendously important. Thus, having just given birth, a mother could decide for herself whether she would give her baby a chance to live. If an adult did not agree with something, he could go and find another camp. Our need for privacy is invaded in an unprecedented way by modern social media, resulting in reputation damage.

An example: in 2006, an evening in Café De Pulp in the Dutch town of Bussum got a bit out of hand. Two middle-aged ladies, dressed in tiger-print blouses, enjoyed the attentions of a small group of merry students, whereupon one of the women allowed herself to be lovingly pawed by one of the lads. He was called Oliver and the reason he involved one of the women in a quickie was a fifty-euro bet. So what, you might say, everyone is free to do as he or she pleases, as long as it is within the boundaries of the law. Alas. A reprobate got it into his head to digitally record the moment of sexual relaxation between the woman and the student *in flagrante delicto,* or just as Oliver put into the practice the ancient art of biting the cake, and as the woman lowered herself onto his naked spindly legs as if onto an old bicycle. Again, nothing amiss of course, until the photos popped up on an internet page with a reference to the woman's place of work, a school in Bussum. The internet exploded. The woman was publicly pilloried, with severe consequences for her job and relationship. Obviously the question immediately arose as to who was 'guilty' of subjecting the poor woman to such humiliation. Had she brought the problems on herself? Could the excitable students be blamed, or the sites on which the photos were posted? Or maybe it was the general public who viewed the images in large numbers and with an avid revulsion that proved them culpable? The fact is that the reputation of the woman will be an issue for years to come, even though legally she has done nothing wrong.

A few years ago, the American blogger Hunter Moore, twenty-five at the time, was knifed in his shoulder by a young woman in San Francisco. Moore had put up nude photos of her on his blog without her permission. Moore posted many nude photos of strangers on a daily basis, some of which were sent in by the strangers themselves, but more often by ex-lovers, attention-seekers or jealous partners. Alongside these photos Moore – who enjoyed the popularity of a cult-figure because he also posted raunchy images and films about himself on his site – put up the email exchanges, furious reactions and death threats that his posts provoked. Taking legal action against Moore was difficult at the time, as he only posted so-called third party content, which meant he was not responsible for it. His pages caused a great deal of distress. Women saw photos of themselves pop up, some people no longer dared to leave the house. But however despicable Moore's conduct may have been, the general public beat a path to his site in massive numbers. Each month he attracted thirty million visitors, netting him approximately ten thousand dollars.

He brushed aside the criticism that he was taking advantage of other people's suffering. In an interview he said: 'It comes down to, you're fucking stupid and I'm making money of your mistakes?' In 2015, a judge found Moore guilty of committing crimes 'in an effort to obtain nude images of people against their will'. His sentence was a minimum of two years and a maximum of seven years in prison, plus a fine of half a million dollars.

'Revenge porn', 'cyber rape' and 'hate porn' are different terms for pornographic images that have been posted on the internet without the knowledge or consent of the people involved. The aim is humiliation or social damage. Often, it is ex-boyfriends who, following the break-up of a relationship, post explicit images online in revenge. Research by a revenge porn site shows that one in ten ex-partners are believed to have threatened publicising private erotic photos. Ninety per cent of the victims of hate porn are female and almost half of these women have been harassed online by people known or unknown to them after the photos were posted. In Great Britain alone, there are alleged to be twenty internet sites that post private pornographic images.

Revenge porn can have enormous consequences. Victims lose their jobs, suffer from depression and find themselves socially isolated. In some cases hate porn can lead to suicide, as in the case of fifteen-year old Amanda Todd. She was talked into flaunting her breasts to a webcam, whereupon these images were distributed amongst her classmates and friends. The person alleged to have persuaded her to do this was a Dutchman by the name of Audin C.; he is said to have blackmailed forty young women over their nude photos. Amanda Todd committed suicide. Seventeen-year old Júlia Rebecca from Brazil took her own life after a sex video of herself and two other minors was posted online. With our primitive brain, we cannot envisage the consequences of our lover taking a nude picture of us on their smartphone.

In many countries, posting revenge porn is now prohibited by law. Censorship (what? censorship!) of particular media could also be a solution. It ought to be easier to remove items from the internet. If girls who are the victims of hate porn were able to wipe these kinds of images with one click, this would be a step in the right direction. Google and Bing: are you reading this?

Digital detox

The world modern humans live in is ever more virtual thanks to the power of media such as books, radio, television and the internet – all new from an evolutionary perspective. The digital revolution is in full swing, and this can lead to our brain increasingly (a) responding to cues that are not real, or (b) failing to respond to cues that are real. We worry about the fate of characters in soap operas as if they were our real friends. We subconsciously copy the behaviour of celebrities, even when that is not very good for us. And our brain confuses porn with real sex. Sometimes we forget there is a real world out there, with good conversations, real friendships and satisfying romantic relationships. The dangers reaching us through TV screens and the internet raise our stress levels and we experience them as if we were personally involved. An attack in a faraway city hits us as if we had witnessed it ourselves.

Humans have cluttered their environment with modern media to such a degree that, as Marshall McLuhan noted, there is no escape from it. How do we turn modern media into match and, going one step further, how can we harness the media and the fact that we live in an increasingly virtual world to tackle the problems surrounding us? How do we prevent the physical world being taken over by a virtual one, creating an environment in which people no longer need to leave their homes, can no longer form normal relationships and can only find sexual satisfaction via the internet (already a major problem in Japan)? If humans were to stop reproducing altogether, all the other living creatures on the planet would probably jump for joy, but we believe there are other options. After all, there are still plenty of people who do not watch TV, are not addicted to Facebook and are impervious to the omnipresence of the media.

Can we cut ourselves off from social media for a while and start living a real life again? Last year Simone Engelen, a young Utrecht-based photographer, invited writers and other artists to organise her life for one day. Simone asked them to provide scripts for a brief, alternative existence, a new life that she would capture in photos. Someone made her go out for a meal with a mirror, another wanted her to be the centre of attention at an exuberant gay party. One author asked if she would spend the day travelling around with his elderly mother, while someone else came up with the idea of Simone spending twenty-four hours alone in a white, windowless room without any books, newspapers, art, photographs, Wi-Fi, computers, telephones or other forms of human expression. The assignment was to spend twenty-four hours focusing solely on herself. Simone saw it through to the end in total solitude. She was so affected by the experiment that she signed up that same night for a ten-day meditation course which involved abstaining from any form of communication. Her aim was complete digital detox.

The human desire to retreat and repose goes back to prehistory. Our ancestors would retreat into the forest, withdrawing from daily life for a while. A digital detox could be the modern variant. One of the authors of this book had a British colleague who, in the days before Wi-Fi, was so addicted to his computer that on Friday afternoons he would unplug it and post the cable to himself. He would have to wait till Monday, when it landed on his doormat, to resume his virtual life.

The Screen-Free Week (formerly TV Turnoff Week and Digital Detox Week) is an annual event that encourages people to turn their electronic equipment off and 'turn their lives on'. There are all kinds of initiatives that promote unplugging, either temporarily or permanently. Many businesses are beginning to switch on to the idea that sending and receiving emails should be confined to office hours, and certainly not during weekends. There are also smartphone apps that show how much time you have spent on social media.

The power of the imagination

Modern media includes powerful tools to change our lifestyles, to turn mismatch into match. 'Each mismatch has its match' as Dutch football legend Johan Cruyff might have put it. Medical professionals all over the world are using media to promote healthy behaviour.

This includes computer programs showing what someone will look like in twenty years' time if they carry on smoking or drinking, encouraging them to be active in front of the TV with Nintendo Wii, or sharing how many kilometres they have jogged on Facebook. Media can also be deployed to open our eyes to the harmful effects of environmental pollution (as did Al Gore's film *An Inconvenient Truth*, mentioned in the previous chapter, which alerted many people to global warming).

Worldwide studies have demonstrated that reading fiction has a universally positive effect. Prisoners who read books, for instance, develop greater empathy and are less likely to reoffend. The authorities in the Italian region of Calabria recently introduced a rule that prisoners could reduce their time in prison by reading books. Each book they read would bring forward their release date by three days, up to a maximum of forty-eight days a year. In the UK, numerous prison reading groups have been set up, and a pilot project is underway in Flanders involving reading clubs made up partly of detainees and partly of ordinary members of the public. They discuss novels together, facilitating the prisoners' return to society. That's why it is so distressing that current governments across Europe, as part of the cuts, have decided to abolish prison libraries.

Perhaps literature can even provide the key to world peace. Canadian and American research shows that people are better able to put themselves in someone else's shoes when they read novels and poetry. A reader sees the world from the protagonist's point of view and this increases their ability to empathise. If it were up to us, hooligans, soldiers, terrorists and gang members would be given free books and reading courses. We could even drop novels over conflict zones. Books, not bombs!

CHAPTER 10

The mismatch test

In this book, we have discussed and analysed the concept of mismatch in great detail. Mismatch occurs when our brain and body have not adapted well to our environment. This has consequences for our behaviour and for the choices we make in our life with regard to nutrition, education, sexual behaviour, politics and work. And this in turn can affect our chances of survival, prosperity and successful procreation.

Our brain and body are perfectly adapted to the environment of prehistory, the period between two and a half million to ten thousand years ago, when humans roamed the savannah in small hunter-gatherer groups. All kinds of traces from this time are still evident in our body and behaviour. We have argued that the most radical changes occurred after the introduction of agriculture, when humans began to settle first in villages and then in towns and cities. From that point on, humans began to interfere actively in their environment. And this resulted in mismatch. After that, the industrial revolution, and more recently the digital revolution we are right in the middle of, have altered our lives from those of our ancestors in key areas. For instance, our diets have changed, the way we travel is different, we tend not to communicate face to face with each other and our work and private lives are often separate. Through our choices we have also fundamentally changed the living environment of other animal and plant species. At times this may mean we save other creatures from certain death, but more often our lifestyle is a contributory factor to other animal and plant species ceasing to exist.

Mismatch can have adverse consequences for our well-being and physical and mental state. We can safely assume that not every person is affected by mismatch to the same degree. This can be related to where on the planet you live. In some areas the environment has changed so fundamentally that it provokes behaviour that no longer works to the reproductive advantage of humans. Countries such as Singapore and Hong Kong are extremely prosperous and yet hardly any children are born there. In other parts of the world the natural environment has become so polluted (through our own actions) that no healthy children are born. Mismatch can also occur to a lesser or greater degree as a result of the choices people themselves make. Some people become addicted to drugs or internet pornography, which takes over their entire lives. Others choose to go on a diet that resembles that of our ancestors. Some only travel to work in their cars, while others jump on their bikes.

Our supposition is that the better someone's live is matched, the greater the chance of a happy and healthy life. Conversely, we think that the more severely someone's life is mismatched, the greater the risks for health and well-being.

At the end of the book, we would like to give you a few final ideas to ponder. Agriculture and the subsequent industrial and digital revolutions have changed our lives fundamentally. These upheavals seem to happen in ever quicker succession. The transformation from farmer to factory worker took our ancestors some ten thousand years, but in less than a hundred years labourers have become digital knowledge workers. What does the future have in store for us? Suppose that, just like you, many readers take what they have learned from this book to heart and offset the many mismatches in their lives and in their direct environment. What will be the result? We will become healthier, happier, older and will eventually have more children, we will do better in school, at work and therefore in the status race ... just as has happened throughout our entire evolution. Will this not immediately invoke the next round of mismatches? Is mismatch commensurate with evolution itself? A mutation that gives evolutionary advantage will do so until this trait has spread throughout the entire population. Then it is back to square one with your slightly longer neck. We have to run to stay in the same place, like the Red Queen in *Alice in Wonderland*.

As historian Yuval Noah Harari argues about our culture in *Sapiens*, what is a luxury for one generation is a necessity for the next. When the mobile phone first appeared, many – we, too, confess – said there was no need for it. And look where we are now. Can you do without it? Do you consider your mobile phone a luxury or a necessity? Our health care is running at full tilt to keep the elderly moving with new hips, transplants, drugs and operations, but despite these interventions, many of these people have mentally long given up the ghost. Our bodies are getting older, while our brains and immune systems barely keep up. Alzheimer's, dementia; you will look in vain for these diseases amongst present-day hunter-gatherers. Are our bodies equipped to live to a hundred or more? Are we, however large our brain, able to escape mismatch?

Our large brain is a primeval organ that has been formed over millions of years of evolution to solve primeval problems. We do not have a brain that can assess climate change properly, or spot the danger of a dictator in charge of a full arsenal of nuclear weapons. And when we do see it, it is too late. One strike is literally enough to send humans towards extinction (just ask the dinosaurs).

And yet neutralising mismatch does not necessarily have to lead to new mismatches. We are above all a social animal species, with an enormous set of creative brains, so it must be possible to create

315

collectively a society in which we are all happier, healthier and more prosperous. This discussion has been held as long as we have had language, but it is only now, partly thanks to evolutionary psychology, that the insights and knowledge are coming together to get to the bottom of the mechanisms.

We do not have the ultimate solution at the ready either, but it will be something along the lines of social sharing, encouraging empathy, and transforming the pursuit of status into contentment. Sufficient wealth for all, with 'sufficient' as the irresistible cue!

Perhaps we can put our money where our mouth is and tackle one of our primal traits – that of sticking our heads in the sand where the future is concerned. According to futurologists, we are on the eve of a genetic revolution which will enable humans to alter themselves genetically. Artificial selection instead of natural selection. That is quite something. By directly intervening in our genome we would thus be able to shake off various mismatches. We could disable the genes that are responsible for diseases such as obesity or cancer, depression or fear of flying. We can choose partners on the basis of their gene compatibility. We can start to fiddle with the DNA of extinct animal species such as the dodo or the Costa Rican toad and give them another chance.

But the biological revolution might very well lead to new mismatches and new misfortune. What will happen when parents are able to choose the gender and character of their children? Will we not all have children of the same sex and with the same personality, resulting in fierce competition between them? How does someone deal with risk if he knows he can be cloned? Will he respond in the same way to threats and danger? How do people react to leaders who have grown in height using growth hormones or athletes who topple world records with gene doping? And why would you try your hardest at school or work if you can give your intelligence a boost with a small genetic procedure? Which of our primal traits will be enhanced through DNA manipulation anyhow? That of the warrior, trader, or social animal? We can use insights from evolutionary psychology and literature (from *Brave New World* to *1984*) to work out which ethical questions are coming at us and in which direction we would collectively like to send them. But something we can predict is that they will be right under our noses before we know it.

That is why, at the end, we would like to offer you something to help you understand yourself better. We will do this in two ways. First, we will give you a list of instantly executable maxims to match your life. See what you can do with them, how easy they are to do and what is in it for you. Call it the Ten Mismatch Commandments, your religious primordial brain will respond well to such cues:

Thou shalt breastfeed
Thou shalt invest sufficiently in thine youngest offspring
Thou shalt play outside more often
Thou shalt eat more fruits and nuts
Thou shalt not manage but lead
Thou shalt recycle thy own waste
Thou shalt not idolise celebrities
Thou shalt look each other in the eyes more frequently
Thou shalt read more books and read aloud more to thy children
Thou shalt ingest more scientific knowledge

And to conclude we offer you a test in which you, the reader, can participate, to see how well your lifestyle and habits match the life to which your body and brain have been optimally adapted. We have come up with some questions, each with two reply options. The left options indicate a match, the right mismatch. So, the more items you tick in the right-hand column, the greater the mismatch. Please note that this is tongue-in-cheek and not a science-based test.

Yet it is interesting to us that you as a reader will find out to what extent mismatch may affect your life. We differentiate various areas of mismatch, including youth, nutrition, housing, relationships, work, politics, religion and nature. Add up all the items you have ticked in the right-hand column. A high score in any one particular area does not mean you need to leg it to the doctor, psychologist or dietician. See it as a cue, or food for thought. Like this entire book, for that matter.

The Mismatch Test

Youth

Were you born in hospital?	No	Yes
Were you breastfed?	Yes	No
Did you live in a city or in the country as a youngster?	Country	City
Are your parents divorced?	No	Yes
Do you have two or more elder brothers and/or sisters?	No	Yes
Did you play outside a lot as a child?	Yes	No
Did you regularly go camping with your parents?	Yes	No

Housing

Do you live near a green area (park, forest)?	Yes	No
Do you live in a flat?	No	Yes
Do you have many plants at home?	Yes	No
Do you buy mainly fresh products?	Yes	No
Do you have a fireplace at home?	Yes	No
Does being in one place for any length of time make you feel restless?	Yes	No

Relationships

Are you in an exclusive relationship?	Yes	No
Did you (or your partner) take the pill when you met?	No	Yes
Do you have children?	Yes	No
Do you have more than two children?	No	Yes
Do you watch pornography?	No	Yes
Is your partner older than you?		
– For men	No	Yes
– For women	Yes	No

Nutrition

Do you like barbecues?	Yes	No
Have you ever hunted?	Yes	No
Have you ever fished?	Yes	No
Do you eat meat with every meal?	No	Yes
Do you eat a lot fruits and nuts?	Yes	No
Are you able to resist a bag of crisps easily?	Yes	No
If you have to choose between beer and wine, what do you choose?	Wine	Beer

Habits

Do you take a lot of exercise?	Yes	No
Do you drink a lot of alcohol?	No	Yes
Do you smoke?	No	Yes
Do you consider yourself overweight?	No	Yes
Do you use a sunbed?	No	Yes

Work

Do you travel to work by car?	No	Yes
Is your work stressful?	No	Yes
Are you self-employed?	Yes	No

Do you gossip much at work?	Yes	No
Are you sensitive to authority?	No	Yes
Do you enjoy meetings?	No	Yes
Does money make you happy?	No	Yes

Politics and Religion

Are you superstitious?	Yes	No
Do you (in the main) vote left or right?	Left	Right
Does politics need a strong leader?	Yes	No
Are you a monarchist?	No	Yes
Do you belong to a church?	No	Yes

The environment

Do you enjoy watching nature documentaries?	Yes	No
Are you a member of an environmental organisation?	Yes	No
Do you recycle at home?	Yes	No
Do you drive an electric car?	Yes	No

Media

Are there days you do not use social media?	Yes	No
Do you have friends you only know through Facebook or Twitter?	No	Yes
Did you meet your partner online?	No	Yes
Do you often take selfies?	No	Yes
Does the news keep you awake at night?	No	Yes

Result

First and foremost and for the sake of completeness: this test has not been scientifically validated and is purely meant for light-hearted insight. Tick your answers, and add up 0 points for each answer in the left-hand column and 1 point for each answer in the right-hand column. Work out your total score.

0–10 points

You have little mismatch in your life. Going back to ancestral times would be no problem for you. But watch out that you do not miss out on too many technical and social trends.

11–20 points

You are reasonably stable. Modern time presents few problems for you, but nor are you entirely impervious to mismatch.

21–30

Things are going moderately well. You adapt to the demands of modern time, but this carries a great risk. Mismatch is lying in wait.

31–40

You are clearly in the danger zone and have an increased risk of mismatch. Be careful that because of all the technological and cultural developments, you do not lose your way.

More than 41

Mismatch has you in its clutches and the modern era is clearly taking you for a ride. Take care that you do not jeopardise your happiness and well-being. Read this book once more. Give it as a present to other people you suspect are experiencing mismatch.

Acknowledgements

Humans have a uniquely human and instinctive ability to show gratitude (the reciprocity principle referred to earlier). We yield to this instinct wholeheartedly and thank everybody who has helped with the writing of this book. We would like to single out Mascha Lammes, Hannie van Hooff, Miranda Bruinzeel, Ruud Hollander, Inge Fraters, Jean-Paul Keulen, Stef Menken, Cécile Koekkoek, Bert Natter, Jeroen van Baaren, Anna Brinkman, Hannah Moore, Toon van de Put, Sander Pinkse, the VU University of Amsterdam, Merijn Hollestelle, Willemijn Lindhout, Joost Nijsen, Trisha de Lang, everyone who works at Podium, our publishers; and our children.

Ronald Giphart and Mark van Vugt

Bibliography

Chapter 1: The mismatch vision

Boyd, R., & Silk, J., *How Humans Evolved*, Norton, 2009

Buss, D. M., *The Handbook of Evolutionary Psychology*, Wiley, 2015

Darwin, C., *On the Origin of Species*, Murray, 1859

Darwin, C., *The Descent of Man*, Murray, 1871

Dawkins, R., *The Selfish Gene*, Oxford University Press, 1976

Diamond, J., *The World Until Yesterday*, Penguin, 2013

Dunbar, R., Barrett, L., & Lycett, J., *Evolutionary Psychology: A Beginner's Guide*, 2007

Gigerenzer, G., *Gut Feelings: The Intelligence of the Unconscious*, Penguin, 2007

Hahn-Holbrook, J., & Haselton, M., 'Is postpartum depression a disease of modern civilization?', *Current Directions in Psychological Science, 23*, 2014

Harari, Y. N., *Sapiens: A Brief History of Humankind*, Random House, 2014

Marlowe, F., *The Hadza*, University of California Press, 2010

Pinker, S., *The Blank Slate: The Denial of Human Nature*, Penguin Classics, 2002

Richerson, P., & Boyd, R., *Not by Genes Alone: How Culture Transformed Human Evolution*, University of Chicago Press, 2004

Robertson, B. A., Rehage, J., & Sih, A., 'Ecological novelty and the emergence of evolutionary traps', *Trends in Ecology and Evolution, 28*, 2013

Schmitt, D. P., & Pilcher, J. J., 'Evaluating evidence of psychological adaptation: How do we know one when we see one', *Current Directions in Psychological Science, 15*, 2004

Stringer, C., *Lone Survivors: How We Came to Be the Only Humans on Earth*, St. Martin's Griffin, 2013

Stulp, G., Pollet, T. V., & Buunk, B. P., 'A curvilinear effect of height on reproductive success in males', *Behavioral Ecology and Sociobiology, 66*, 2012

Tomasello, M., *The Cultural Origins of Human Cognition*, Harvard University Press, 2009

Vugt, M. van, & Ahuja, A., *Selected: Why Some People Lead, Why Others Follow, and Why it Matters,* Profile, 2010

Wrangham, R., *Catching Fire: How Cooking Made Us Human*, Basic Books, 2010

Chapter 2: Old bodies, modern diseases

Buss, D. M., *The Handbook of Evolutionary Psychology*, Wiley, 2015

Campbell, A., 'Staying alive: Evolution, culture, and women's intrasexual aggression', *Behavioral and Brain Sciences*, 22, 1999

Cronin, H., *The Ant and the Peacock: Altruism and Sexual Selection from Darwin to Today*, Cambridge University Press, 1993

Durkheim, E., *Suicide: A Study in Sociology*, The Free Press, 1897

Geary, D. C., *Male, Female: The Evolution of Human Sex Differences*, American Psychological Association, 1998

Gilbert, P., *Depression: The Evolution of Powerlessness*, Routledge, 2014

Gluckman, P., & Hanson, M., *Mismatch: Why Our World No Longer Fits Our Bodies*, Oxford University Press, 2006

Gray, P., 'Play as a foundation for hunter-gatherer existence', *American Journal of Play*, 2009

Hahn-Holbrook, J., & Haselton, M., 'Is postpartum depression a disease of modern civilization?', *Current Directions in Psychological Science*, 23, 2014

Hawkes, K., 'Human longevity: the grandmother effect', *Nature*, 428, 2004

Hertwig, R., Davis, J., & Sulloway, F., 'Parental investment: How an equity motive can produce inequality', *Psychological Bulletin*, 128, 2002

Hough, R. A., *Captain James Cook: A Biography*, W. W. Norton & Company, 2013

Hrdy, S. B., *Mothers and Others: The Evolutionary Origins of Mutual Understanding*, Harvard University Press, 2009

Kaplan, H., Hill, K., Lancaster, J., & Hurtado, M., 'A theory of human life history evolution: Diet, intelligence and longevity', *Evolutionary Anthropology*, 2000

Kuipers, R., *Fatty Acids in Human Evolution: Contributions to Evolutionary Medicine*, University of Groningen, 2012

Kuipers, R. S., *The Primeval Diet: The Way to Get as Healthy as our Ancestors (Het oerdieet. De manier om oergezond te worden)*, Bert Bakker, 2014

Moss, M., *Salt, Sugar, Fat*, Random House, 2014

Nelissen, M., *Darwin's Glasses (De bril van Darwin)*, Lannoo, 2015

Nesse, R., & Williams, G. C., *Why We Get Sick: The New Science of Darwinian Medicine*, Vintage, 1996

Pagel, M., & Bodmer, W., 'A naked ape would have fewer parasites', *Proceedings of Royal Society-B*, 270, 2003

Wal, A. van der, Krabbendam, A. C., Schade, H., & Vugt, M. van, 'Do natural landscapes reduce future discounting in humans?', *Proceedings of the Royal Society-B*, 280, 2013

Wansink, B., & Chandon, P., 'Slim by design: redirecting the accidental drivers of overeating', *Journal of Consumer Psychology*, 24, 2014.

Warinner, C., *Debunking the Meat Myth* [TEDx talk], 2014

Chapter 3: A crazy little thing called love

Bellis, M., Hughes, K., Hughes, S., & Ashton, J., 'Measuring paternal discrepancy and its public health consequences', *Journal of Epidemiology and Community Health*, 59, 2005

Bryson, B., *Shakespeare: The World as a Stage*, Atlas Books, 2009

Buss, D. M., 'Sex differences in human mate preferences: evolutionary hypotheses tested in 37 cultures', *Behavioral and Brain Sciences*, 1989

Chapais, B., 'Monogamy, strongly bonded groups and the evolution of human social structure', *Evolutionary Anthropology*, 22, 2013

Cook, M., *A Brief History of the Human Race*, W. W. Norton & Company, 2005

Dunbar, R., Baron, R., Frangou, A., Pearce, E., van Leeuwen, E., Stow, J., Partridge, G., Macdonald, I., Barra, V., & Vugt, M. van, 'Social laughter is correlated with an elevated pain threshold', *Proceedings of the Royal Society-B*, 2011

Eibl-Eibesfeldt, I., *Ethology: The Biology of Behavior*, Holt, Rinehart and Winston, 1970

Ellis, B., 'Timing of pubertal maturation in girls: An integrated life history approach', *Psychological Bulletin*, *130*, 2004

Fisher, Helen E., *Anatomy of Love: a Natural History of Mating, Marriage, and Divorce*, Simon & Schuster, 1992

Friedl, E., 'Sex the Invisible', *American Anthropologist*, *96*, 1994

Gildersleeve, K., Haselton, M. G., & Fales, M. R., 'Do women's mate preferences change across the ovulatory cycle? A meta-analytic review', *Psychological Bulletin*, 2014

Giphart, R., *The Feast of Love (Het feest der liefde)*, Balans, 1995

Gomes, C., & Boesch, C., 'Wild chimpanzees exchange meat for sex on a long-term basis', *PLoS ONE*, 2009

Griskevicius, V., et al., 'Environmental contingency in life history strategies: the influence of mortality and socio-economic status on reproductive timing', *Journal of Personality and Social Psychology*, *100*, 2011

Harcourt, A. H., Harvey, P. H., Larson, S. G., & Short, R. V., 'Testis weight, body weight and breeding system in primates', *Nature*, 1981

Hooff, J. C. van, Crawford, H., Vugt, M. van, 'The wandering mind of men: ERP evidence for gender differences in attention bias towards attractive opposite sex faces', *Social Cognitive and Affective Neuroscience*, 2010

Hrdy, S. B., *Mothers and Others: The Evolutionary Origins of Mutual Understanding*, Harvard University Press, 2009

Iredale, W., Vugt, M. van, & Dunbar, R., 'Showing off in humans: Male generosity as mate signal', *Evolutionary Psychology*, *6*, 2008

Marlowe, F., *The Hadza*, University of California Press, 2010

Meston, C., & Buss, D., 'Why humans have sex', *Archives of Sexual Behavior*, *36*, 2007

Meston, C., & Buss, D., *Why Women Have Sex*, St. Martin's Griffin, 2010

Rikowski, A., & Grammer, K., 'Human body odour, symmetry, and attractiveness', *Proceedings of the Royal Society-B*, 1999

Roberts, S. C., et al., 'Partner choice, relationship satisfaction, and oral conception: the congruency hypothesis', *Psychological Science*, *25*, 2014

Thornhill, R., & Palmer, C. T., *A Natural History of Rape: Biological Basis of Sexual Coercion*, The MIT Press, 2000

Vugt, M. van, & Iredale, W., 'Men behaving nicely: Public goods as peacock tails', *British Journal of Psychology*, *104*, 2013

Chapter 4: This isn't working

Backer, C. de, *Like Belgian Chocolate for the Universal Mind: Interpersonal and Media Gossip from an Evolutionary Perspective*, Ghent University, 2005

Backer, C. de, *Gossiping, Why Gossiping is Healthy (Roddelen, waarom roddelen gezond is)*, Unieboek, 2006

Bruin, E. de, *Meetings? Don't Do It! (Vergaderen? Niet doen!)*, Atlas, 2014

Buunk, B. P., *Primitive Drives in the Workplace (Oerdriften op de werkvloer)*, Prometheus, 2010

Colarelli, S. M., *No Best Way: An Evolutionary Perspective on Human Resource Management*, Praeger, 2003

Colarelli, S. M., & Arvey, R., *The Biological Foundations of Organizational Behavior*, Chicago University Press, 2015

Cosmides, L., Barrett, C., & Tooby, J., 'Adaptive specializations, social exchange, and the evolution of human intelligence', *PNAS*, *107*, 2010

Diamond, J., *Guns, Germs and Steel: The Fates of Human Societies*, Random House, 1998

Dunbar, R., *Grooming, Gossip and the Evolution of Language*, Faber & Faber, 1996

Dunbar, R., 'Gossip in evolutionary perspective', *Review of General Psychology*, *8*, 2004

Frumin, I., et al., 'A social chemosignaling function for human handshaking', *E-Life*, 2015

Hawkes, K., et al., 'Why hunter-gatherers work', *Current Anthropology*, *34*, 1993

Johnson, D. D. P., Price, M. E., & Vugt, M. van, 'Darwin's invisible hand: Market competition, evolution and the firm', *Journal of Economic Behavior and Organization*, *90*, 2013

Lee, R. B., *The !Kung San: Men, Women and Work in a Foraging Society*, Cambridge University Press, 1979

Nicholson, N., *Managing the Human Animal*, Texere, 2010

Ronay, R. D., Greenaway, K., Anicich, E., & Galinsky, A., 'The path to glory is paved with hierarchy', *Current Directions in Psychological Science*, *23*, 2012

Saad, G., *Evolutionary Psychology in the Business Sciences*, Springer, 2011

Sahlins, M., *Stone Age Economics*, Aldine Transaction, 1972

Sapolsky, R. M., *Why Zebras Don't Get Ulcers: A Guide to Stress, Stress-Related Diseases, and Coping*, Freeman, 1994

Smith, A., *The Wealth of Nations*, 1776

Vugt, M. van, & Ahuja, A., *Naturally selected: why some people lead, why other follow, and why it matters*, Profile, 2010

Weggeman, M., *Managing Professionals?Don't Do It! (Leidinggeven aan professionals? Niet doen!)*, Scriptum, 2012

Wilkinson, R., *The Spirit Level: Why More Equal Societies Almost Always Do Better*, Bloomsbury Press, 2011

Chapter 5: Follow the leader

Bass, B. M., *Bass and Stogdill's Handbook of Leadership: Theory, Research, and Managerial Applications* (3rd edition), Free Press, 1990

Blaker, N. M., Rompa, I., Dessing, I. H., Vriend, A. F., Herschberg, C., & Vugt, M. van, 'The height leadership advantage in men and women: Testing evolutionary psychology predictions about the perceptions of tall leaders', *Group Processes and Intergroup Relations*, *16*, 2013

Boehm, C., *Hierarchy in the Forest*, Harvard University Press, 1999

Hare, R., & Babiak, P., *Snakes in Suits*, HarperBusiness, 2007

Hartog, D. N. den, House, R. J., Hanges, P. J., Ruiz-Quintanilla, S. A., & Dorfman, P. W., 'Culture specific and cross-culturally generalizable implicit leadership theories: Are attributes of charismatic/transformational leadership universally endorsed?', *The Leadership Quarterly*, 1999

Henrich, J., & Gil-White, F., 'The evolution of prestige: Freely conferred deference as a mechanism for enhancing the benefits of cultural transmission', *Evolution and Human Behavior*, *22*, 2001

Herman, S., & Boer, J., *Eat, Drink, Sleep (Eten, drinken, slapen)*, De Librije, 2010

Hollander, E. P., 'The essential interdependence of leadership and followership', *Current Directions in Psychological Science*, *1*, 1992

Kiatpongsan, S., & Norton, M. I., 'How much (more) should CEOs make?', *Perspectives on Psychological Science*, *9*, 2014

King, A., Johnson, D., & Vugt, M. van, 'The origins and evolution of leadership', *Current Biology*, 2009

Luyendijk, J., *Swimming with Sharks: My Journey into the World of Bankers,* Guardian Books and Faber & Faber, 2015

Meindl, J. R., Ehrlich, S. B., & Dukerich, J. M., 'The romance of leadership', *Administrative Science Quarterly*, *30*, 1985

Pinker, S., *The Blank Slate*, Penguin Classics, 2002

Rueden, C. von, Gurven, M., & Kaplan, H., 'Why do men seek status? Fitness payoffs to dominance and prestige', *Proceedings of the Royal Society-B*, *283*, 2010.

Rueden, C. von, & Vugt, van M., 'Leadership in small scale societies: Some implications for theory, research and practice', *The Leadership Quarterly*, 2015

Sjøgren, K., 'The boss, not the workload causes workplace depression', *ScienceNordic*, 2013

Smith, J. E., Gavrilets, S., Borgerhoff Mulder, M., Hooper, P. L., El Moulden, C., Nettle, D., Hauert, C., Hill, K., Perry, S., Pusey, A. E., Van Vugt, M., & Smith, E. A., 'Leadership in mammalian societies: Emergence, distribution, power, and pay-off', *Trends in Ecology and Evolution*, 2015

Vugt, M. van, & Ahuja, A., *Naturally Selected: The evolutionary science of leadership*, Harper, 2011

Vugt, M. van, & Grabo, A. E., 'The many faces of leadership: An evolutionary psychology approach', *Current Directions in Psychological Science*, 2015

Vugt, M. van, Hogan, R., & Kaiser, R., 'Leadership, followership, and evolution: Some lessons from the past', *American Psychologist*, *63*, 2008

Vugt, M. van, Jepson, S. F., Hart, C. M., & De Cremer, D., 'Autocratic leadership in social dilemmas: A threat to group stability', *Journal of Experimental Social Psychology*, *40*, 2004

Vugt, M. van, & Wildschut, M., *Authority: The Science of Power, Authority and Leadership (Gezag: De wetenschap van macht, gezag, en leiderschap)*, A. W. Bruna, 2013

Waal, F. de, *Chimpanzee Politics: Power and Sex Among Apes*, John Hopkins University Press, 2007

Wilkinson, R., *The Spirit Level: Why More Equal Societies Almost Always Do Better*, Bloomsbury Press, 2011

Chapter 6: The god paradox

Atran, S., *In Gods We Trust: The Evolutionary Landscape of Religion*, Oxford University Press, 2002

Atran, S., 'Genesis of suicide terrorism', *Science*, 2003

Boyer, P., 'Being human: Religion: bound to believe', *Nature*, *455*, 2008

Dawkins, R., *The God Delusion*, Random House, 2009

Diamond, J., *The World Until Yesterday*, Penguin, 2013

Johnson, D., 'God's punishment and public goods', *Human Nature*, *16*, 2005

Norenzayan, A., *Big Gods: How Religion Transformed Cooperation and Conflict*, Princeton University Press, 2013

Norenzayan, A., & Sharrif, A. F., 'The origin and evolution of religious prosociality', *Science*, *322*, 2008

O'Gorman, R. O., Henrich, J., & Vugt, M. van, 'Constraining free-riding in public goods games: Designated solitary punishers can sustain human cooperation', *Proceedings of Royal Society-B*, 276, 2008

Poloma, M., & Pendleton, B., 'The effects of prayer and prayer experiences on measures of general wellbeing', *Journal of Psychology and Theology*, 19, 1911

Skinner, B., 'Superstition in the pigeon', *Journal of Experimental Psychology*, 1948

Sosis, R., 'Why aren't we all Hutterites?', *Human Nature*, 2003

Vugt, M. van, & Ahuja, A., *Selected: Why Some People Lead, Why Others Follow, and Why it Matters*, Profile Books, 2010

Chapter 7: War, what is it good for?

Alexander, R. D., *The Biology of Moral Systems*, Aldine, 1987

Batson, D., et al., 'Is empathy-induced helping due to self-other merging?', *Journal of Personality and Social Psychology*, 73, 1997

Campbell, A., 'Staying alive: Evolution, culture, and women's intrasexual aggression', *Behavioral and Brain Sciences*, 22, 1999

Carneiro, R. L., 'A theory of the origin of the state', *Science*, 169, 1970

Chagnon, N. A., 'Life histories, blood revenge, and warfare in a tribal population', *Science*, 239, 1988

Dallek, R., *Flawed Giant: Lyndon Johnson and His Times*, Oxford University Press, 1999

Dreu, C. de, et al., 'Oxytocin and parochial altruism', *Science*, 2011

Goldstein, J., *War and Gender*, Cambridge University Press, 2003

Keegan, J., *A History of Warfare*, Random House, 1994

Keeley, L., *War Before Civilization*, Oxford University Press, 1996

Mathew, S., & Boyd, R., 'Punishment sustains large-scale cooperation in prestate warfare', *PNAS*, 2011

Mazur, A., & Booth, A., 'Testosterone and dominance in men', *Behavioral and Brain Sciences*, 1998

McDonald, M., Navarrete, C., & Vugt, M. van, 'The male warrior hypothesis', *Philosophical Transactions of the Royal Society*, 367, 2012

Moore, M., *Bowling for Columbine* [film], 2002

Navarrete, C. D., Fessler, D. M. T., & Eng, S. J., 'Elevated ethnocentrism in the first trimester of pregnancy', *Evolution and Human Behavior*, 28, 2007

Pinker, S., *The Better Angels of our Nature: Why Violence Has Declined*, Viking Books, 2011

Rusch, H., Leunissen, J. M., & Vugt, M. van, 'Historical and experimental evidence of sexual selection for war heroism', *Evolution and Human Behavior*, 2015

Sander, H., *Liberators take Liberties (Befreier und Befreite)* [film], 1992

Sidanius, J., & Pratto, F., *Social Dominance: An Intergroup Theory of Social Hierarchy and Oppression*, Cambridge University Press, 1999

Taylor, S. E., et al., 'Biobehavioral responses to stress in females: Tend-and-befriend, not fight-or-flight', *Psychological Review*, 107, 2000

Tooby, J., & Cosmides, L., 'The evolution of war and its cognitive foundations', *Institute for Evolutionary Studies Technical Report*, 1988

Vugt, M. van, 'Sex differences in intergroup competition, aggression, and warfare: The male warrior hypothesis', *Annals of the New York Academy of Sciences*, 1167, 2009

Vugt, M. van, De Cremer, D., & Janssen, D. P., 'Gender differences in cooperation and competition: The male-warrior hypothesis', *Current Directions in Psychological Science*, 18, 2007

Vugt, M. van, & Spisak, B., 'Sex differences in the emergence of leadership during competitions within and between groups', *Current Directions in Psychological Science, 19,* 2008

Wilson, M., & Wrangham, R., 'Intergroup relations in chimpanzees', *Annual Review of Anthropology, 32,* 2003

Chapter 8: Waiting for the ice to melt

Burnstein, E., Crandall, C., & Kitayama, S., 'Some neo-Darwinian decision rules for altruism: Weighing the cues for inclusive fitness as a function of the biological importance of the decision', *Journal of Personality and Social Psychology, 67,* 1994

Buunk, A. P., & Vugt, M. van, *Applying Social Psychology: From Problems to Solutions,* Sage, 2013

Cialdini, R. B., 'Crafting normative messages to protect the environment', *Current Directions in Psychological Science, 12,* 2003

Coyne, J., *Why Evolution is True,* Penguin, 2009

Cuddy, A. J. C., & Doherty, K. T., 'OPOWER: Increasing energy efficiency through normative influence', *Harvard Business School Case N9-911-16,* 2010

Diamond, J., *Collapse: How Societies Choose to Fail or Succeed,* Allen Lane, 2005

Ehrlich, P., *Human Natures: Genes, Cultures, and the Human Prospect,* Island Press, 2000

Frank, R., *Choosing the Right Pond: Human Behavior and the Quest for Status,* Oxford University Press, 1985

Gifford, R., 'The dragons of inaction: Psychological barriers that limit climate change mitigation and adaptation', *American Psychologist, 66,* 2011

Goldstein, N. J., Cialdini, R. B., &

Griskevicius, V., 'A room with a viewpoint: Using social norms to motivate environmental conservation in hotels', *Journal of Consumer Research,* 2008

Griskevicius, V., Tybur, J. M., & Van den Bergh, B., 'Going green to be seen: Status, reputation, and conspicuous conservation', *Journal of Personality and Social Psychology, 98,* 2010

Hardin, G., 'Tragedy of the commons', *Science, 162,* 1968

Kaplan, R., & Kaplan, S., *The Experience of Nature: A Psychological Perspective,* Cambridge University Press, 1989

Lange, P. van, Balliet, D., Parks, C., & Vugt, M. van, *Social Dilemmas,* Oxford University Press, 2014

Low, B. S., 'Behavioral ecology of conservation in traditional societies', *Human Nature, 7,* 1996

Martin, P. S., *Twilight of the Mammoths: Ice Age Extinctions and the Rewilding of America,* University of California Press, 2005

OECD, *OECD Environmental Outlook to 2050,* OECD Publishing, 2012

Ostrom, E., *Governing the Commons,* Cambridge University Press, 1990

Penn, D., & Mysterud, J., *Evolutionary Perspectives on Environmental Problems,* Transaction Publishers, 2007

Quammen, D., *The Song of the Dodo: Island Biogeography in an Age of Extinctions,* Olympus, 2009

Rousseau, J.-J., *Discourse on the origin of inequality (Discours sur l'origine et les fondements de l'inégalité parmi less hommes)* translated by Franklin Philip, Oxford University Press, 1994

Ulrich, R. S., 'View through a window may influence recovery from surgery', *Science, 4647,* 1984

Vugt, M. van, 'Community identification moderating the impact of financial incentives in a natural social dilemma: Water conservation',

Personality and Social Psychology Bulletin, 25, 2001

Vugt, M. van, 'Averting the tragedy of the commons: Using social psychological science to protect the environment', *Current Directions in Psychological Science*, 18, 2009

Vugt, M. van, & Griskevicius, V., 'UNESCO World Social Science Report: Changing global environments. Going green: Using evolutionary psychology to foster sustainable lifestyles', *ISSC*, 2013

Vugt, M. van, Griskevicius, V., & Schultz, P. W., 'Naturally green: Harnessing stone age psychological biases to foster environmental behavior', *Social Issue and Policy Review*, 8, 2014

Wal, A. van der, Krabbendam, A. C., Schade, H., & Vugt, M. van, 'Do Natural Landscapes Reduce Future Discounting in Humans?', *Proceedings of the Royal Society-B*, 280, 2013

Wilson, E. O., *The Creation: An Appeal to Save Life on Earth*, Norton, 2006

Chapter 9: Virtual reality

Barrett, D., *Supernormal Stimuli: How Primal Urges Overran their Evolutionary Purpose*, W. W. Norton & Company, 2010

Boyd, B., *On the Origin of Stories: Evolution, Cognition, and Fiction*, Harvard University Press, 2009

Backer, C. de, *Like Belgian Chocolate for the Universal Mind: Interpersonal and Media Gossip from an Evolutionary Perspective*, Ghent University, 2005

Carr, N., *The Shallows*, W. W. Norton & Company, 2010

Chatwin, B., *The Songlines*, Cape, 1987

Dunbar, R., *Grooming, Gossip and the Evolution of Language*, Faber & Faber, 1996

Dunbar, R., 'Gossip in evolutionary perspective', *Review of General Psychology*, 8, 2004

Eggers, D., *The Circle*, Penguin Books, 2014

Gigerenzer, G., *Gut Feelings: The Intelligence of the Unconscious*, Penguin, 2007

Haselton, M. G., & Buss, D. M., 'Error management theory: A new perspective on biases in cross-sex mind reading', *Journal of Personality and Social Psychology*, 78, 2000

Henrich, J., *The Secret of our Success: How Culture is Driving Human Evolution, Domesticating Our Species, and Making Us Smarter*, Princeton University Press, 2015

Kanazawa, S., 'Evolutionary psychology and intelligence research', *American Psychologist*, 2010

Kidd, D. C., & Castano, E., 'Reading literary fiction improves theory of mind', *Science*, 342, 2013

Mcluhan, M., *The Medium is the Message*, John Wiley, 1967

Miller, G., *The Mating Mind: How Sexual Selection Shaped the Evolution of Human Nature*, Doubleday, 2000

Mithen, S., *After the Ice*, Weidenfeld & Nicolson, 2003

Mithen, S., *The Singing Neanderthals: The Origins of Music, Language, Mind and Body*, Weidenfeld & Nicholson, 2005

Pinker, S., *The Language Instinct*, Penguin, 1995

Pinker, S., *The Village Effect: Why Face-to-face Contact Matters*, Atlantic Books, 2014

Robinson, A., *The Story of Writing*, Thames & Hudson, 2007

Rojek, C., *Celebrity. The Wiley Blackwell Encyclopedia of Consumption and Consumer Studies*, Wiley-Blackwell, 2015

Sontag, S., *On Photography*, FsG, 1977

Chapter 10: The mismatch test

Harari, Y. N., *Sapiens: A Brief History of Humankind*, Random House, 2014

Index

INDEX